T0205442

Studies in Computational Intelligence

Volume 673

Series editor

Janusz Kacprzyk, Polish Academy of Sciences, Warsaw, Poland
e-mail: kacprzyk@ibspan.waw.pl

About this Series

The series "Studies in Computational Intelligence" (SCI) publishes new developments and advances in the various areas of computational intelligence—quickly and with a high quality. The intent is to cover the theory, applications, and design methods of computational intelligence, as embedded in the fields of engineering, computer science, physics and life sciences, as well as the methodologies behind them. The series contains monographs, lecture notes and edited volumes in computational intelligence spanning the areas of neural networks, connectionist systems, genetic algorithms, evolutionary computation, artificial intelligence, cellular automata, self-organizing systems, soft computing, fuzzy systems, and hybrid intelligent systems. Of particular value to both the contributors and the readership are the short publication timeframe and the worldwide distribution, which enable both wide and rapid dissemination of research output.

More information about this series at http://www.springer.com/series/7092

Wojciech Wieczorek

Grammatical Inference

Algorithms, Routines and Applications

 Springer

Wojciech Wieczorek
Institute of Computer Science
University of Silesia Faculty of Computer
   Science and Materials Science
Sosnowiec
Poland

ISSN 1860-949X   ISSN 1860-9503   (electronic)
Studies in Computational Intelligence
ISBN 978-3-319-83589-1   ISBN 978-3-319-46801-3   (eBook)
DOI 10.1007/978-3-319-46801-3

Printed on acid-free paper

This Springer imprint is published by Springer Nature
The registered company is Springer International Publishing AG
The registered company address is: Gewerbestrasse 11, 6330 Cham, Switzerland

Preface

Grammatical inference, the main topic of this book, is a scientific area that lies at the intersection of multiple fields. Researchers from computational linguistics, pattern recognition, machine learning, computational biology, formal learning theory, and many others have their own contribution. Therefore, it is not surprising that the topic has also a few other names such as grammar learning, automata inference, grammar identification, or grammar induction. To simplify the location of present contribution, we can divide all books relevant to grammatical inference into three groups: theoretical, practical, and applicable. In greater part this book is practical, though one can also find the elements of learning theory, combinatorics on words, the theory of automata and formal languages, plus some reference to real-life problems.

The purpose of this book is to present old and modern methods of grammatical inference from the perspective of practitioners. To this end, the Python programming language has been chosen as the way of presenting all the methods. Included listings can be directly used by the paste-and-copy manner to other programs, thus students, academic researchers, and programmers should find this book as the valuable source of ready recipes and as an inspiration for their further development.

A few issues should be mentioned regarding this book: an inspiration to write it, a key for the selection of described methods, arguments for selecting Python as an implementation language, typographical notions, and where the reader can send any critical remarks about the content of the book (subject–matter, listings etc.).

There is a treasured book entitled "Numerical recipes in C", in which along with the description of selected numerical methods, listings in C language are provided. The reader can copy and paste the fragments of the electronic version of the book in order to produce executable programs. Such an approach is very useful. We can find an idea that lies behind a method and immediately put it into practice. It is a guiding principle that accompanied writing the present book.

For the selection of methods, we try to keep balance between importance and complexity. It means that we introduced concepts and algorithms which are essential to the GI practice and theory, but omitted that are too complicated or too long to present them as a ready-to-use code. Thanks to that, the longest program included in the book is no more than a few pages long.

As far as the implementation language is concerned, the following requirements had to be taken into account: simplicity, availability, the property of being firmly established, and allowing the use of wide range of libraries. Python and FSharp programming languages were good candidates. We decided to choose IronPython (an implementation of Python) mainly due to its integration with the optimization modeling language. We use a monospaced (fixed-pitch) font for the listings of programs, while the main text is written using a proportional font. In listings, Python keywords are in bold.

The following persons have helped the author in preparing the final version of this book by giving valuable advice. I would like to thank (in alphabetical order): Prof. Z.J. Czech (Silesian University of Technology), Dr. P. Juszczuk, Ph.D. student A. Nowakowski, Dr. R. Skinderowicz, and Ph.D. student L. Strak (University of Silesia).

Sosnowiec, Poland   Wojciech Wieczorek
2016

Contents

Acronyms

CFG	Context-free grammar
CGT	Combinatorial game theory
CNF	Chomsky normal form
CNF	Conjunctive normal form
CSP	Constraint satisfaction problem
DFA	Deterministic finite automaton
DNF	Disjunctive normal form
EDSM	Evidence driven state merging
GA	Genetic algorithm
GI	Grammatical inference
GNF	Greibach normal form
ILP	Integer linear programming
LP	Linear programming
MDL	Minimum description length
MILP	Mixed integer linear programming
NFA	Non-deterministic finite automaton
NLP	Non-linear programming
NP	Non-deterministic polynomial time
OGF	Ordinary generating function
OML	Optimization modeling language
PTA	Prefix tree acceptor
RPNI	Regular positive and negative inference
SAT	Boolean satisfiability problem
TSP	Traveling salesman problem
XML	Extensible markup language

Chapter 1
Introduction

1.1 The Problem and Its Various Formulations

Let us start with the presentation of how many variants of a grammatical inference problem we may be faced with. Informally, we are given a sequence of words and the task is to find a rule that lies behind it. Different models and goals are given by response to the following questions. Is the sequence finite or infinite? Does the sequence contain only examples (positive words) or also counter-examples (negative words)? Is the sequence of the form: all positive and negative words up to a certain length n? What is meant by the rule: are we satisfied with regular acceptor, context-free grammar, context-sensitive grammar, or other tool? Among all the rules that match the input, should the obtained one be of a minimum size?

1.1.1 Mathematical Versus Computer Science Perspectives

The main division of GI models comes from the size of a sequence. When it is infinite, we deal with mathematical identification in the limit. The setting of this model is that of on-line, incremental learning. After each new example, the learner (the algorithm) must return some hypothesis (an automaton or a CFG). Identification is achieved when the learner returns a correct answer and does not change its decision afterwards. With respect to this model the following results have been achieved: (a) if we are given examples and counter-examples of the language to be identified (*learning from informant*), and each individual word is sure of appearing, then at some point the inductive machine will return the correct hypothesis; (b) if we are given only the examples of the target (*learning from text*), then identification is impossible for any super-finite class of languages, i.e., a class containing all finite languages and at least one infinite language. In this book, however, we only consider the situation when the input is finite, which can be called a computer science perspective. We are going to describe algorithms the part of which base on examples only, and the others base

© Springer International Publishing AG 2017
W. Wieczorek, *Grammatical Inference*, Studies in Computational Intelligence 673,
DOI 10.1007/978-3-319-46801-3_1

Fig. 1.1 A DFA accepting
aa*bb*cc*

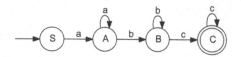

on both examples and counter-examples. Sometimes we will demand the smallest possible form of an output, but every so often we will be satisfied with an output that is just consistent with the input. Occasionally, an algorithm gives the collection of results that gradually represent the degree of generalization of the input.

1.1.2 Different Kinds of Output

The next point that should be made is that how important it is to pinpoint the kind of a target. Consider the set of examples: {abc, aabbcc, aaabbbccc}. If a solution is being sought in the class of regular languages, then one possible guess is presented in Fig. 1.1. This automaton matches every word starting with one or more as, followed by one or more bs, and followed by one or more cs. If a solution is being sought in the class of context-free languages, then one of possible answers is the following grammar:

$$S \to A\,B\,C \quad A \to a \mid a\,A\,B$$
$$B \to b \qquad C \to c \mid C\,C$$

It is clearly seen that the language accepted by this CFG is $\{a^m b^m c^n : m, n \geq 1\}$. Finally, if a solution is being sought in the class of context-sensitive languages, then even more precise conjecture can be made:

$$S \to a\,B\,c \quad a\,A \to a\,a \quad b\,A \to A\,b$$
$$B \to A\,b\,B\,c \quad B\,c \to b\,c$$

Now, the language accepted by this grammar is $\{a^m b^m c^m : m \geq 1\}$. It is worth emphasizing that from the above-mentioned three various acceptors, the third one can not be described as a CFG or a DFA, and the second one can not be described as a DFA (or an NFA). In the book we will only consider the class of regular and context-free languages. The class of context-sensitive languages is very rarely an object of investigations in the GI literature. Such identification is thought to be a very hard task. Moreover, the decision problem that asks whether a certain word belongs to the language of a given context-sensitive grammar, is PSPACE-complete.

It should be noted that in the absence of counter-examples, we are at risk of over-generalization. An illustration that suggests itself is the regular expression $(a + b + c)^*$, which represents all words over the alphabet $\{a, b, c\}$, as a guess for the examples from the previous paragraph. In the present book, two approaches to this problem have been proposed. First, the collection of different hypotheses is suggested, and it is up to a user to select the most promising one. Second, the minimum description length (MDL) principle is applied, in which not only the size of an output matters but also the amount of information that is needed to encode every example.

1.1.3 Representing Languages

Many definitions specific to particular methods are put in relevant sections. Herein, we give definitions that will help to understand the precise formulation of our GI problem and its complexity. Naturally, we skip conventional definitions and notation from set theory, mathematical logic, and discrete structures, in areas that are on undergraduate courses.

Definition 1.1 Σ will be a finite nonempty set, the *alphabet*. A *word* (or sometimes *string*) is a finite sequence of symbols chosen from an alphabet. For a word w, we denote by $|w|$ the length of w. The *empty word* λ is the word with zero occurrences of symbols. Sometimes, to be coherent with external notation, we write *epsilon* instead of the empty word. Let x and y be words. Then xy denotes the *catenation* of x and y, that is, the word formed by making a copy of x and following it by a copy of y. We denote, as usual by Σ^*, the set of all words over Σ and by Σ^+ the set $\Sigma^* - \{\lambda\}$. A word w is called a *prefix* (resp. a *suffix*) of a word u if there is a word x such that $u = wx$ (resp. $u = xw$). The prefix or suffix is *proper* if $x \neq \lambda$. Let $X, Y \subset \Sigma^*$. The *catenation* (or product) of X and Y is the set $XY = \{xy \mid x \in X, y \in Y\}$. In particular, we define

$$X^0 = \{\lambda\}, \quad X^{n+1} = X^n X \ (n \geq 0), \quad X^{\leq n} = \bigcup_{i=0}^{n} X^i. \tag{1.1}$$

Definition 1.2 Given two words $u = u_1 u_2 \ldots u_n$ and $v = v_1 v_2 \ldots v_m$, $u < v$ according to *lexicographic order* if $u = v_1 \ldots v_n$ ($n < m$) or if $u_i < v_i$ for the minimal i where $u_i \neq v_i$. The *quasi-lexicographic order* on the Σ^* over an ordered alphabet Σ orders words firstly by length, so that the empty word comes first, and then within words of fixed length n, by lexicographic order on Σ^n.

To simplify the representations for languages (i.e., sets of words), we define the notion of regular expressions, finite-state automata, and context-free grammars over alphabet Σ as follows.

Definition 1.3 The set of *regular expressions* (regexes) over Σ will be the set of words R such that

1. $\emptyset \in R$ which represents the empty set.
2. $\Sigma \subseteq R$, each element a of the alphabet represents language $\{a\}$.

3. If r_A and r_B are regexes representing languages A and B, respectively, then $(r_A + r_B) \in R$, $(r_A r_B) \in R$, $(r_A^*) \in R$ representing $A \cup B$, AB, A^*, respectively, where the symbols $(,), +, ^*$ are not in Σ.

We will freely omit unnecessary parentheses from regexes assuming that catenation has higher priority than $+$ and * has higher priority than catenation. If $r \in R$ represents language A, we will write $L(r) = A$.

The above definition of regular expressions has a minor drawback. When we write, for instance, abc, it is unknown whether the word denotes itself or one of the following regexes: ((ab)c), (a(bc)). In the present book, however, it will be always clear from the context.

Definition 1.4 A *non-deterministic finite automaton* (NFA) is defined by a quintuple $(Q, \Sigma, \delta, s, F)$, where Q is the finite set of states, Σ is the input alphabet, $\delta\colon Q \times \Sigma \to 2^Q$ is the transition function, $s \in Q$ is the initial state, and $F \subseteq Q$ is the set of final states. When there is no transition at all from a given state on a given input symbol, the proper value of δ is $\emptyset$, the empty set. We extend δ to words over Σ by the following inductive definition:

1. $\delta(q, \lambda) = \{q\}$,
2. $\delta(q, ax) = \bigcup_{r \in \delta(q,a)} \delta(r, x)$, where $x \in \Sigma^*$, and $a \in \Sigma$.

Having extended the transition function δ from the domain $Q \times \Sigma$ to the domain $Q \times \Sigma^*$, we can formally define that a word x belongs to the language accepted by automaton A, $x \in L(A)$, if and only if $\delta(s, x) \cap F \neq \emptyset$.

Definition 1.5 If in a NFA A for every pair $(q, a) \in Q \times \Sigma$, the transition function holds $|\delta(q, a)| \leq 1$, then A is called a *deterministic finite automaton* (DFA).

The *transition diagram* of an NFA (as well as DFA) is an alternative way to represent an NFA. For $A = (Q, \Sigma, \delta, s, F)$, the transition diagram of A is a labeled digraph (V, E) satisfying

$$V = Q, \quad E = \{q \xrightarrow{a} p\colon p \in \delta(q, a)\},$$

where $q \xrightarrow{a} p$ denotes an edge (q, p) with label a. Usually, final states are depicted as double circled terms. A word x over Σ is accepted by A if there is a labeled path from s to a state in F such that this path spells out the word x.

Definition 1.6 A *context-free grammar* (CFG) is defined by a quadruple $G = (V, \Sigma, P, S)$, where V is an alphabet of *variables* (or sometimes *non-terminal symbols*), Σ is an alphabet of *terminal symbols* such that $V \cap \Sigma = \emptyset$, P is a finite set of *production rules* of the form $A \to \alpha$ for $A \in V$ and $\alpha \in (V \cup \Sigma)^*$, and S is a special non-terminal symbol called the *start symbol*. For the sake of simplicity, we will write $A \to \alpha_1 \mid \alpha_2 \mid \ldots \mid \alpha_k$ instead of $A \to \alpha_1$, $A \to \alpha_1, \ldots, A \to \alpha_k$. We call a word $x \in (V \cup \Sigma)^*$ a *sentential form*. Let u, v be two words in $(V \cup \Sigma)^*$ and $A \in V$. Then, we write $uAv \Rightarrow uxv$, if $A \to x$ is a rule in P. That is, we can substitute word

x for symbol A in a sentential form if $A \rightarrow x$ is a rule in P. We call this rewriting a *derivation*. For any two sentential forms x and y, we write $x \Rightarrow^* y$, if there exists a sequence $x = x_0, x_1, \ldots, x_n = y$ of sentential forms such that $x_i \Rightarrow x_{i+1}$ for all $i = 0, 1, \ldots, n - 1$. The language $L(G)$ generated by G is the set of all words over Σ that are generated by G; that is, $L(G) = \{x \in \Sigma^* \mid S \Rightarrow^* x\}$.

Definition 1.7 A *sample* S over Σ will be an ordered pair $S = (S_+, S_-)$ where S_+, S_- are finite subsets of Σ^* and $S_+ \cap S_- = \emptyset$. S_+ will be called the *positive part of S (examples)*, and S_- the *negative part of S (counter-examples)*. Let A be one of the following acceptors: a regular expression, a DFA, an NFA, or a CFG. We call A *consistent* (or *compatible*) with a sample $S = (S_+, S_-)$ if and only if $S_+ \subseteq L(A)$ and $S_- \cap L(A) = \emptyset$.

1.1.4   Complexity Issues

The consecutive theorems are very important in view of computational intractability of GI in its greater part. These results will be given without proofs because of the scope and nature of the present book. Readers who are interested in details of this may get familiar with works listed at the end of this chapter (Sect. 1.4).

Theorem 1.1 *Let Σ be an alphabet, S be a sample over Σ, and k be a positive integer. Determining whether there is a k-state DFA consistent with S is NP-complete.*

It should be noted, however, that the problem can be solved in polynomial time if $S = \Sigma^{\leq n}$ for some n. The easy task is also finding such a DFA A that $L(A) = S_+$ and A has the minimum number of states.

Theorem 1.2 *The decision version of the problem of converting a DFA to a minimal NFA is PSPACE-complete.*

Let us answer the question of whether NFA induction is a harder problem than DFA induction. The search space for the automata induction problem can be assessed by the number of automata with a fixed number of states. It has been shown that the number of pairwise non-isomorphic minimal k-state DFAs over a c-letter alphabet is of order $k2^{k-1}k^{(c-1)k}$, while the number of NFAs on k states over a c-letter alphabet such that every state is reachable from the start state is of order 2^{ck^2}. Thus, switching from determinism to non-determinism increases the search space enormously. On the other hand, for $c, k \geq 2$, there are at least 2^{k-2} distinct languages $L \subseteq \Sigma^*$ such that: (a) L can be accepted by an NFA with k states; and (b) the minimal DFA accepting L has 2^k states. It is difficult to resist the conclusion that—despite its hardness—NFA induction is extremely important and deserves exhaustive research.

Theorem 1.3 *Let Σ be an alphabet, S be a sample over Σ, and k be a positive integer. Determining whether there is a regular expression r over Σ, that has k or fewer occurrences of symbols from Σ, and such that r is consistent with S, is NP-complete.*

Interestingly, the problem remains NP-complete even if r is required to be star-free (contains no "*" operations).

When moving up from the realm of regular languages to the realm of context-free languages we are faced with additional difficulties. The construction of a minimal cover-grammar seems to be intractable, specially in view of the following facts: (a) there is no polynomial-time algorithm for obtaining the smallest context-free grammar that generates exactly one given word (unless $P = NP$); (b) context-free grammar equivalence and even equivalence between a context-free grammar and a regular expression are undecidable; (c) testing equivalence of context-free grammars generating finite sets needs exponential time; (d) the grammar can be exponentially smaller than any word in the language.

1.1.5 Summary

Let Σ be an alphabet, and k, n be positive integers. We will denote by $\mathbf{I}$ (input) a sample $S = (S_+, S_-)$ over Σ for which either S_+, $S_- \neq \emptyset$ or $S_+ \neq \emptyset$ and $S_- = \emptyset$. We write $\mathbf{O}$ (output) for one of the following acceptors: a regex, a DFA, an NFA, or a CFG. The notation $|\mathbf{O}| = k$ means that, respectively, a regex has k occurrences of symbols from Σ, a DFA (or NFA) has k states, and a CFG has k variables.

From now on, every GI task considered in the book will be put into one of the three formulations:

1. For a given $\mathbf{I}$ find a consistent $\mathbf{O}$.
2. For given $\mathbf{I}$ and $k > 0$ find a consistent $\mathbf{O}$ such that $|\mathbf{O}| \leq k$.
3. For a given $\mathbf{I}$ find a consistent $\mathbf{O}$ such that $|\mathbf{O}|$ is minimal.

But not all combinations will be taken into account. For example, in the book there is no algorithm for finding a minimum regex or for finding a CFG for examples only. For technical or algorithmic reasons, sometimes we will demand S not including λ.

If we are given a huge $\mathbf{I}$ with positive and negative words along with an algorithm $\mathscr{A}$ that works for positive words only,[1] then the following incremental procedure can be applied. First, sort S_+ in a quasi-lexicographic order to $(s_1, s_2, \ldots, s_n)$. Next, set $j = 2$ (or a higher value), infer $\mathscr{A}(I = \{s_1, \ldots, s_{j-1}\})$, and check whether $L(\mathscr{A}(I)) \cap S_- = \emptyset$. If no counter-example is accepted, find the smallest k from the indexes $\{j, j+1, \ldots, n\}$ for which s_k is not accepted by $\mathscr{A}(I)$, and set $j = k + 1$. Take the set $I = I \cup \{s_k\}$ as an input for the next inference. Pursue this incremental procedure until $j = n + 1$. Note that the proposed technique is a heuristic way of reducing the running time of the inference process, with no guarantee of getting the correct solution.

---

[1]This technique can also be used for algorithms that work on $S = (S_+, S_-)$, provided that their computational complexity primarily depends on the size of S_+.

1.2 Assessing Algorithms' Performance

Suppose that we are given a sample S from a certain domain. How would we evaluate a GI method or compare more such methods. First, a proper measure should be chosen. Naturally, it depends on a domain's characteristic. Sometimes only precision would be sufficient, but in other cases relying only on a single specific measure—without calculating any general measure of the quality of binary classifications (like Matthews correlation coefficient or others)—would be misleading.

After selecting the measure of error or quality, we have to choose between three basic scenarios. (1) The target language is known, and in case of regular languages we simply check the equivalence between minimal DFAs; for context-free languages we are forced to generate the first n words in the quasi-lexicographic order from the hypothesis and from the target, then check the equality of two sets. Random sampling is also an option for verifying whether two grammars describe the same language. When the target is unknown we may: (2) random split S into two subsets, the training and the test set (T&T) or (3) apply K-fold cross-validation (CV), and then use a selected statistical test. Statisticians encourage us to choose T&T if $|S| > 1\,000$ and CV otherwise. In this section, McNemar's test for a scenario (2) and 5×2 CV t test for a scenario (3), are proposed.

1.2.1 Measuring Classifier Performance

Selecting one measure that will describe some phenomenon in a population is crucial. In the context of an average salary, take for example the arithmetic mean and the median from the sample: $55,000$, $59,000$, $68,000$, $88,000$, $89,000$, and $3,120,000$. Our mean is 579833.33. But it seems that the median,[2] $78,000$ is better suited for estimating our data. Such examples can be multiplied. When we are going to undertake an examination of binary classification efficiency for selected real biological or medical data, making the mistake to classify a positive object as a negative one, may cause more damage than incorrect putting an object to the group of positives; take examples—the ill, counter-examples—the healthy, for good illustration. Then, recall (also called sensitivity, or a true positive rate) would be more suited measure than accuracy, as recall quantifies the avoiding of false negatives.

By binary classification we mean mapping a word to one out of two classes by means of inferred context-free grammars and finite-state automata. The acceptance of a word by a grammar (resp. an automaton) means that the word is thought to belong to the positive class. If a word, in turn, is not accepted by a grammar (resp. an automaton), then it is thought to belong to the negative class. For binary classification a variety of measures has been proposed. Most of them base on rates that are shown

---

[2]The median is the number separating the higher half of a data sample, a population, or a probability distribution, from the lower half. If there is an even number of observations, then there is no single middle value; the median is then usually defined to be the mean of the two middle values.

Table 1.1 Confusion matrix for two classes

	True class		
Predicted class	Positive	Negative	Total
Positive	tp: true positive	fp: false positive	p'
Negative	fn: false negative	tn: true negative	n'
Total	p	n	N

in Table 1.1. There are four possible cases. For a positive object (an example), if the prediction is also positive, this is a *true positive*; if the prediction is negative for a positive object, this is a *false negative*. For a negative object, if the prediction is also negative, we have a *true negative*, and we have a *false positive* if we predict a negative object as positive. Different measures appropriate in particular settings are given below:

- accuracy, $\text{ACC} = (tp + tn)/N$,
- error, $\text{ERR} = 1 - \text{ACC}$,
- balanced accuracy, $\text{BAR} = (tp/p + tn/n)/2$,
- balanced error, $\text{BER} = 1 - \text{BAR}$,
- precision, $\text{P} = tp/p'$,
- true positive rate (eqv. with recall, sensitivity), $\text{TPR} = tp/p$,
- specificity, $\text{SPC} = tn/n$,
- F-measure, $\text{F1} = 2 \times \text{P} \times \text{TPR}/(\text{P} + \text{TPR})$,
- Matthews correlation coefficient, $\text{MCC} = (tp \times tn - fp \times fn)/\sqrt{p \times p' \times n \times n'}$.

If the class distribution is not uniform among the classes in a sample (e.g. the set of examples forms a minority class), so as to avoid inflated measurement, the balanced accuracy should be applied. It has been shown that BAR is equivalent to the AUC (the area under the ROC curve) score in case of the binary (with 0/1 classes) classification task. So BAR is considered as primary choice if a sample is highly imbalanced.

1.2.2 McNemar's Test

Given a training set and a test set, we use two algorithms to infer two acceptors (classifiers) on the training set and test them on the test set and compute their errors. The following natural numbers have to be determined:

- e_1: number of words misclassified by 1 but not 2,
- e_2: number of words misclassified by 2 but not 1.

Under the null hypothesis that the algorithms have the same error rate, we expect $e_1 = e_2$. We have the chi-square statistic with one degree of freedom

$$\frac{(|e_1 - e_2| - 1)^2}{e_1 + e_2} \sim \chi_1^2 \tag{1.2}$$

and McNemar's test rejects the hypothesis that the two algorithms have the same error rate at significance level α if this value is greater than $\chi_{\alpha,1}^2$. For $\alpha = 0.05$, $\chi_{0.05,1}^2 = 3.84$.

1.2.3 5 × 2 Cross-Validated Paired t Test

This test uses training and test sets of equal size. We divide the dataset S randomly into two parts, $x_1^{(1)}$ and $x_1^{(2)}$, which gives our first pair of training and test sets. Then we swap the role of the two halves and get the second pair: $x_1^{(2)}$ for training and $x_1^{(1)}$ for testing. This is the first fold; $x_i^{(j)}$ denotes the j-th half of the i-th fold. To get the second fold, we shuffle S randomly and divide this new fold into two, $x_2^{(1)}$ and $x_2^{(2)}$. We then swap these two halves to get another pair. We do this for three more folds.

Let $p_i^{(j)}$ be the difference between the error rates (for the test set) of the two classifiers (obtained from the training set) on fold $j = 1, 2$ of replication $i = 1, \ldots, 5$. The average on replication i is $\overline{p}_i = (p_i^{(1)} + p_i^{(2)})/2$, and the estimated variance is $s_i^2 = (p_i^{(1)} - \overline{p}_i)^2 + (p_i^{(2)} - \overline{p}_i)^2$. The null hypothesis states that the two algorithms have the same error rate. We have t statistic with five degrees of freedom

$$\frac{p_1^{(1)}}{\sqrt{\sum_{i=1}^{5} s_i^2 / 5}} \sim t_5. \tag{1.3}$$

The 5×2 CV paired t test rejects the hypothesis that the two algorithms have the same error rate at significance level α if this value is outside the interval $(-t_{\alpha/2, 5}, t_{\alpha/2, 5})$. If significance level equals 0.05, then $t_{0.025, 5} = 2.57$.

1.3 Exemplary Applications

As a concept, grammatical inference is a broad field that covers a wide range of possible applications. We can find scientific research and practical implementations from such fields as:

- Natural language processing: building syntactic parsers, language modeling, morphology, phonology, etc. In particular, one of its founding goals is modeling language acquisition.
- Bioinformatics and biological sequence analysis: modeling automatically RNA and protein families of sequences.

- Structural pattern recognition is a field in which grammatical inference has been active since 1974.
- Program engineering and software testing: verification, testing, specification learning.
- Music: classification, help in creating, data recovering.
- Others: malware detection, document classification, wrapper induction (XML), navigation path analysis, and robotic planning.

Some bibliographical references associated with these topics have been reported in Sect. 1.4. Further in this section, a selected CGT (combinatorial game theory) problem as well as a problem from the domain of bioinformatics are solved as an illustration of GI methods usage.

1.3.1   Peg Solitaire

The following problem has been formulated in one of CGT books:

> Find all words that can be reduced to one peg in one-dimensional Peg Solitaire. (A move is for a peg to jump over an adjacent peg into an empty adjacent space, and remove the jumped-over peg: for instance, $1101 \to 0011 \to 0100$, where 1 represents a peg and 0 an empty space.) Examples of words that can be reduced to one peg are $1, 11, 1101, 110101, 1(10)^k 1$. Georg Gunther, Bert Hartnell and Richard Nowakowski found that for an $n \times 1$ board with one empty space, n must be even and the space must be next but one to the end. If the board is cyclic, the condition is simply n even.

Let us solve this problem using a GI method. Firstly, it should be noticed that leading and trailing 0s are unimportant, so we will only consider words over $\{0, 1\}$ that start and end with 1s. Secondly, for every word of length $n \geq 3$ going outside, i.e., $11dd \cdots 1 \to 100dd \cdots 1$, will always put us to a position that cannot be reduced to one peg. Thus, among all moves that can be made from a word w ($|w| \geq 3$), we do not need to take into account those moves which increase the size of a word (one-dimensional board).

Under the hypothesis that the solution being sought, say H, is a regular language, the idea for solving the problem can be stated in three steps: (1) for some n generate $H \cap \{0, 1\}^{\leq n}$; (2) using the k-tails method (from Sect. 3.2) find a sequence of automata; (3) see if any one of them have passed a verification test.

Step 1

To the previously given remarks we can add another one: H is subset of $L(1(1 + 01 + 001)^*)$. In order to determine all words from H up to a certain length, dynamic programming has been applied, using the fact that for any word with m 1s, a jump leads to a word with $m - 1$ 1s and the length of the word does not increase (for words consisting of three or more digits). The algorithm based on dynamic programming will work if as input we give $I = L(1(1 + 01 + 001)^*) \cap \{0, 1\}^{\leq n}$ sorted by the number of 1 s in ascending order. We can easily find out that $|I| = 1104$ for $n = 12$ and $|I| = 6872$ for $n = 15$. This leads us to the following algorithm.

```python
from FAdo.fa import *
from FAdo.reex import *

def moves(w):
  result = set()
  n = len(w)
  if w[:3] == '110':
    result.add('1' + w[3:])
  for i in xrange(2, n-2):
    if w[i-2:i+1] == '011':
      result.add(w[:i-2] + '100' + w[i+1:])
    if w[i:i+3] == '110':
      result.add(w[:i] + '001' + w[i+3:])
  if w[-3:] == '011':
    result.add(w[:-3] + '1')
  return result

def pegnum(w):
  c = 0
  for i in xrange(len(w)):
    if w[i] == '1':
      c += 1
  return c

def generateExamples(n):
  """Generates all peg words of length <= n
  Input: n in {12, 15}
  Output: the set of examples"""
  rexp = str2regexp("1(1 + 01 + 001)*")
  raut = rexp.toNFA()
  g = EnumNFA(raut)
  numWords = {12: 1104, 15: 6872}[n]
  g.enum(numWords)
  words = sorted(g.Words, \
    cmp = lambda x, y: cmp(pegnum(x), pegnum(y)))
  S_plus = {'1', '11'}
  for i in xrange(4, numWords):
    if moves(words[i]) & S_plus:
      S_plus.add(words[i])
  return S_plus
```

Step 2

The second step is just the invocation of the k-tails algorithm. Because the algorithm outputs the sequence of hypothesis, its invocation has been put into the for-loop structure as we can see in a listing from step 3.

Step 3

The target language is unknown, that is why we have to propose any test for probable correctness of obtaining automata. To this end, we generate two sets of words, namely positive test set (Test_pos) and negative test set (Test_neg). The former contains all words form the set $H \cap \{0, 1\}^{\leq 15}$, the latter contains the remaining words over $\{0, 1\}$ up to the length 15. An automaton is supposed to be correct if it accepts all words from the positive test set and accepts no word from the negative test set.

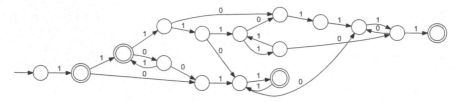

Fig. 1.2 A DFA representing H

```
def allWords(n):
    """Generates all words over {0, 1} up to length n
    Input: an integer n
    Output: all w in (0 + 1)* such that 1 <= |w| <= n"""
    rexp = str2regexp("(0 + 1)(0 + 1)*")
    raut = rexp.toNFA()
    g = EnumNFA(raut)
    g.enum(2**(n+1) - 2)
    return set(g.Words)

Train_pos = generateExamples(12)
Test_pos = generateExamples(15)
Test_neg = allWords(15) - Test_pos

for A in synthesize(Train_pos):
    if all(A.evalWordP(w) for w in Test_pos) \
      and not any(A.evalWordP(w) for w in Test_neg):
        Amin = A.toDFA().minimalHopcroft()
        print Amin.Initial, Amin.Final, Amin.delta
        break
```

The resulting automaton is depicted in Fig. 1.2.

1.3.2 Classification of Proteins

One of the problems studied in bioinformatics is classification of amyloidogenic
hexapeptides. Amyloids are proteins capable of forming fibrils instead of the functional structure of a protein, and are responsible for a group of diseases called amyloidosis, such as Alzheimer's, Huntington's disease, and type II diabetes. Furthermore,
it is believed that short segments of proteins, like hexapeptides consisting of 6-residue
fragments, can be responsible for amyloidogenic properties. Since it is not possible
to experimentally test all such sequences, several computational tools for predicting
amyloid chains have emerged, inter alia, based on physico-chemical properties or
using machine learning approach.

Suppose that we are given two sets, training and test, of hexapeptides:

```
Train_pos = \
['STQIIE', 'STVIIL', 'SDVIIE', 'STVIFE', 'STVIIS', 'STVFIE', \
 'STVIIN', 'WIVIFF', 'YLNWYQ', 'SFQIYA', 'SFFFIQ', 'STFIIE', \
 'GTFFIN', 'ETVIIE', 'SEVIIE', 'YTVIIE', 'STVIIV', 'SMVLFS', \
```

```
 'STVIYE', 'VILLIS', 'SQFYIT', 'SVVIIE', 'STVIII', 'HLVYIM', \
 'IEMIFV', 'FYLLYY', 'FESNFN', 'TTVIIE', 'STVIIF', 'STVIIQ', \
 'IFDFIQ', 'RQVLIF', 'ITVIIE', 'KIVKWD', 'LTVIIE', 'WVFWIG', \
 'SLVIIE', 'STVTIE', 'STVIIE', 'GTFNII', 'VSFEIV', 'GEWTYD', \
 'KLLIYE', 'SGVIIE', 'STVNIE', 'GVNYFL', 'STLIIE', 'GTVLFM', \
 'AGVNYF', 'KVQIIN', 'GTVIIE', 'WTVIIE', 'STNIIE', 'AQFIIS', \
 'SSVIIE', 'KDWSFY', 'STVIIW', 'SMVIIE', 'ALEEYT', 'HYFNIF', \
 'SFLIFL', 'STVIIA', 'DCVNIT', 'NHVTLS', 'EGVLYV', 'VEALYL', \
 'LAVLFL', 'STSIIE', 'STEIIE', 'STVIIY', 'LYQLEN', 'SAVIIE', \
 'VQIVYK', 'SIVIIE', 'HGWLIM', 'STVYIE', 'QLENYC', 'MIENIQ']
Train_neg = \
['KTVIVE', 'FHPSDI', 'FSKDWS', 'STVITE', 'STVDIE', 'FMFFII', \
 'YLEIII', 'STVIDE', 'RMFNII', 'ETWFFG', 'NGKSNF', 'KECLIN', \
 'STVQIE', 'IQVYSR', 'AAELRN', 'EYLKIA', 'KSNFLN', 'DECFFF', \
 'STVPIE', 'YVSGFH', 'EALYLV', 'HIFIIM', 'RVNHVT', 'AEVLAL', \
 'PSDIEV', 'STVIPE', 'DILTYT', 'RETWFF', 'STVIVE', 'KTVIYE', \
 'KLLEIA', 'QPKIVK', 'EECLFL', 'QLQLNI', 'IQRTPK', 'YAELIV', \
 'KAFIIQ', 'GFFYTP', 'HPAENG', 'KTVIIT', 'AARRFF', 'STVIGE', \
 'LSFSKD', 'NIVLIM', 'RLVFID', 'STVSIE', 'LSQPKI', 'RGFFYT', \
 'YQLENY', 'QFNLQF', 'ECFFFE', 'SDLSFS', 'KVEHSD', 'STVMIE', \
 'QAQNQW', 'SSNNFG', 'TFWEIS', 'VTLSQP', 'STVIEE', 'TLKNYI', \
 'LRQIIE', 'STGIIE', 'YTFTIS', 'SLYQLE', 'DADLYL', 'SHLVEA', \
 'SRHPAE', 'KWDRDM', 'FFYTPK', 'STVIQE', 'GMFNIQ', 'HKALFW', \
 'LLWNNQ', 'GSHLVE', 'VTQEFW', 'NIQYQF', 'STMIIE', 'PTEKDE', \
 'TNELYM', 'LIAGFN', 'HAFLII', 'YYTEFT', 'EKNLYL', 'KTVLIE', \
 'FTPTEK', 'STPIIE', 'STVVIE', 'SGFHPS', 'LFGNID', 'SPVIIE', \
 'STVISE', 'EKDEYA', 'RVAFFE', 'FYTPKT', 'PKIQVY', 'DDSLFF', \
 'ERGFFY', 'PTVIIE', 'DIEVDL', 'STIIIE']
Test_pos = \
['FTVIIE', 'HQLIIM', 'ISFLIF', 'GTFFIT', 'YYQNYQ', 'HFVWIA', \
 'NTVIIE', 'SNVIIE', 'MLVLFV', 'YVEYIG', 'STVWIE', 'STVIIM', \
 'EYSNFS', 'SQVIIE', 'SYVIIE', 'FLVHSS', 'NQQNQY', 'QYFNQM', \
 'DTVIIE', 'VTSTFS', 'STVIIT', 'LIFLIV', 'SFVIIE', 'NYVWIV', \
 'NFGAIL', 'STVIID', 'VTVIIE', 'MTVIIE', 'STVLIE', 'LLYYTE', \
 'QTVIIE', 'KLFIIQ', 'ATVIIE', 'LVEALY', 'TYVEYI', 'RVFNIM', \
 'NQFIIS', 'STVEIE']
Test_neg = \
['STDIIE', 'LKNGER', 'KAILFL', 'NYFAIR', 'VKWDRD', 'KENIIF', \
 'WVENYP', 'WYFYIQ', 'VAQLNN', 'DLLKNG', 'HLVEAL', 'TAWYAE', \
 'STAIIE', 'STVGIE', 'ERIEKV', 'EVDLLK', 'STVIIP', 'AINKIQ', \
 'STTIIE', 'TYQIIR', 'MYFFIF', 'TEFTPT', 'NGERIE', 'AENGKS', \
 'ICSLYQ', 'YASEIE', 'VAWLKM', 'NLGPVL', 'RTPKIQ', 'EHSDLS', \
 'AEMEYL', 'NYNTYR', 'TAELIT', 'HTEIIE', 'AEKLFD', 'LAEAIG', \
 'STVIME', 'GERGFF', 'VYSRHP', 'YFQINN', 'SWVIIE', 'KGENFT', \
 'STVIWE', 'STYIIE', 'QTNLYG', 'HYQWNQ', 'IEKVEH', 'KMFFIQ', \
 'ILENIS', 'FFWRFM', 'STVINE', 'STVAIE', 'FLKYFT', 'FGELFE', \
 'STVILE', 'WSFYLL', 'LMSLFG', 'FVNQHL', 'STVIAE', 'KTVIIE', \
 'MYWIIF']
```

The biologist's question is which method, based on decision trees or based on grammatical inference will achieve better accuracy.

As regards decision trees approach, classification and regression trees (CART), a non-parametric decision tree learning technique, has been chosen. For this purpose we took advantage of a scikit-learn's[3] optimized version of the CART algorithm:

---

[3] Scikit-learn provides a range of supervised and unsupervised learning algorithms via a consistent interface in Python. It is licensed under a permissive simplified BSD license. Its web page is http://scikit-learn.org/stable/.

```python
from sklearn import tree
from sklearn.externals.six import StringIO
from functools import partial

Sigma = set(list("NMLKIHWVTSRQYGFEDCAP"))
idx = dict(zip(list(Sigma), range(len(Sigma))))

def findACC(f):
  score = 0
  for w in Test_pos:
    if f(w):
      score += 1
  for w in Test_neg:
    if not f(w):
      score += 1
  if score == 0:
    return 0.0
  else:
    return float(score)/float(len(Test_pos) + len(Test_neg))

def acceptsBy(clf, w):
  return clf.predict([map(lambda c: idx[c], list(w))])[0] == 1

X = []
Y = []
for x in Train_pos:
  X.append(map(lambda c: idx[c], list(x)))
  Y.append(1)
for y in Train_neg:
  X.append(map(lambda c: idx[c], list(y)))
  Y.append(0)
clf = tree.DecisionTreeClassifier()
clf = clf.fit(X, Y)
print findACC(partial(acceptsBy, clf))
```

As for GI approach, it was the induction of a minimal NFA choice, since the sample is not very large.

```python
from functools import partial
from FAdo.common import DFAsymbolUnknown

def acceptsBy(aut, w):
  try:
    return aut.evalWordP(w)
  except DFAsymbolUnknown:
    return False

S_plus, S_minus = set(Train_pos), set(Train_neg)
k = 1
while True:
  print k,
  A = synthesize(S_plus, S_minus, k)
  if A:
    print findACC(partial(acceptsBy, A))
    print A.dotFormat()
    break
  k += 1
```

The resulting automaton is depicted in Fig. 1.3. The ACC scores for the obtained decision tree and NFA are equal to, respectively, 0.616 and 0.677. It is worth emphasizing,

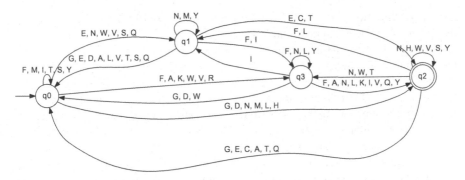

Fig. 1.3 An NFA consistent with the (`Train_pos`, `Train_neg`) sample

though, that McNemar's test does not reject the hypothesis that the two algorithms have the same error rate at significance level 0.05 (the computed statistic was equal to 0.543, so the null hypothesis might be rejected if $\alpha \geq 0.461$).

1.4 Bibliographical Background

Different mathematical models of grammatical inference come from various papers. Language identification in the limit has been defined and studied by Gold (1967). Two other models also dominate the literature. In the *query learning model* by Angluin (1988), the learning algorithm is based on two query instructions relevant to the unknown grammar G: (a) membership—the input is a string w and the output is 'yes' if w is generated by G and 'no' otherwise, (b) equivalence—the input is a grammar G' and the output is 'yes' if G' is equivalent to G and 'no' otherwise. If the answer is 'no', a string w in the symmetric difference of the language $L(G)$ and the language $L(G')$ is returned. In the *probably approximately correct learning model* by Valiant (1984), we assume that random samples are drawn independently from examples and counterexamples. The goal is to minimize the probability of learning an incorrect grammar.

The complexity results devoted to GI from combinatorial perspective were discovered by Gold (1978), for DFAs, and Angluin (1976), for regular expressions. The hardness of NFA minimization was studied by Meyer and Stockmeyer (1972) and Jiang and Ravikumar (1993). The information on the number of k-state DFAs and NFAs is taken from Domaratzki et al. (2002). Some negative results about context-free languages, which also are mentioned in Sect. 1.1.4, come from papers: Charikar et al. (2005), Hunt et al. (1976), and books: Hopcroft et al. (2001), Higuera (2010). An algorithm for finding the minimal DFA consistent with a sample $S = \Sigma^{\leq n}$ was constructed by Trakhtenbrot and Barzdin (1973).

Beside regexes, DFAs, NFAs, and CFGs, there are other string-rewriting tools that can be applied to the grammatical inference domain. Good books for such alternatives are Book and Otto (1993) and Rozenberg and Salomaa (1997). Some positive GI results in this context can be found in Eyraud et al. (2007).

Parsing with context-free grammars is an easy task and is analyzed in many books, for example in Grune and Jacobs (2008). However, context-sensitive membership problem is PSPACE-complete, which was shown by Kuroda (1964). In fact, the problem remains hard even for deterministic context-sensitive grammars.

A semi-incremental method described at the end of Sect. 1.1 was introduced by Dupont (1994). Imada and Nakamura (2009) also applied the similar approach of the learning process with a SAT solver.

Statistical tests given in this chapter were compiled on Alpaydin (2010). Another valuable book, especially when we need to compare two (or more) classifiers on multiple domains, was written by Japkowicz and Shah (2011).

A list of practical applications of grammatical inference can be found in many works; the reader can refer to Bunke and Sanfelieu (1990), de la Higuera (2005), Higuera (2010), and Heinz et al. (2015) as good starting points on this topic. The first exemplary application (peg solitaire) is stated as the 48th unsolved problem in combinatorial games Nowakowsi (1996). This problem was solved by Moore and Eppstein (2003) and we have verified that our automaton is equivalent to the regular expression given by them. The second exemplary application is a hypothetical problem, though, the data are real and come from Maurer-Stroh et al. (2010).

References

Alpaydin E (2010) Introduction to machine learning, 2nd edn. The MIT Press

Angluin D (1976) An application of the theory of computational complexity to the study of inductive inference. PhD thesis, University of California

Angluin D (1988) Queries and concept learning. Mach Learn 2(4):319–342

Book RV, Otto F (1993) String-rewriting systems. Springer, Text and Monographs in Computer Science

Bunke H, Sanfelieu A (eds) (1990) Grammatical inference. World Scientific, pp 237–290

Charikar M, Lehman E, Liu D, Panigrahy R, Prabhakaran M, Sahai A, Shelat A (2005) The smallest grammar problem. IEEE Trans Inf Theory 51(7):2554–2576

de la Higuera C (2005) A bibliographical study of grammatical inference. Pattern Recogn 38(9):1332–1348

de la Higuera C (2010) Grammatical inference: learning automata and grammars. Cambridge University Press, New York, NY, USA

Domaratzki M, Kisman D, Shallit J (2002) On the number of distinct languages accepted by finite automata with n states. J Autom Lang Comb 7:469–486

Dupont P (1994) Regular grammatical inference from positive and negative samples by genetic search: the GIG method. In: Proceedings of 2nd international colloquium on grammatical inference, ICGI '94, Lecture notes in artificial intelligence, vol 862. Springer, pp 236–245

Eyraud R, de la Higuera C, Janodet J (2007) Lars: a learning algorithm for rewriting systems. Mach Learn 66(1):7–31

Gold EM (1967) Language identification in the limit. Inf Control 10:447–474

Gold EM (1978) Complexity of automaton identification from given data. Inf Control 37:302–320

Grune D, Jacobs CJ (2008) Parsing techniques: a practical guide, 2nd edn. Springer

Heinz J, de la Higuera C, van Zaanen M (2015) Grammatical inference for computational linguistics. Synthesis lectures on human language technologies. Morgan & Claypool Publishers

Hopcroft JE, Motwani R, Ullman JD (2001) Introduction to automata theory, languages, and computation, 2nd edn. Addison-Wesley

Hunt HB III, Rosenkrantz DJ, Szymanski TG (1976) On the equivalence, containment, and covering problems for the regular and context-free languages. J Comput Syst Sci 12:222–268

Imada K, Nakamura K (2009) Learning context free grammars by using SAT solvers. In: Proceedings of the 2009 international conference on machine learning and applications, IEEE computer society, pp 267–272

Japkowicz N, Shah M (2011) Evaluating learning algorithms: a classification perspective. Cambridge University Press

Jiang T, Ravikumar B (1993) Minimal NFA problems are hard. SIAM J Comput 22:1117–1141

Kuroda S (1964) Classes of languages and linear-bounded automata. Inf Control 7(2):207–223

Maurer-Stroh S, Debulpaep M, Kuemmerer N, Lopez de la Paz M, Martins IC, Reumers J, Morris KL, Copland A, Serpell L, Serrano L et al (2010) Exploring the sequence determinants of amyloid structure using position-specific scoring matrices. Nat Methods 7(3):237–242

Meyer AR, Stockmeyer LJ (1972) The equivalence problem for regular expressions with squaring requires exponential space. In: Proceedings of the 13th annual symposium on switching and automata theory, pp 125–129

Moore C, Eppstein D (2003) One-dimensional peg solitaire, and duotaire. In: More games of no chance. Cambridge University Press, pp 341–350

Nowakowski RJ (ed) (1996) Games of no chance. Cambridge University Press

Rozenberg G, Salomaa A (eds) (1997) Handbook of formal languages, vol 3. Beyond words. Springer

Trakhtenbrot B, Barzdin Y (1973) Finite automata: behavior and synthesis. North-Holland Publishing Company

Valiant LG (1984) A theory of the learnable. Commun ACM 27:1134–1142

Chapter 2
State Merging Algorithms

2.1 Preliminaries

Before we start analyzing how the state merging algorithms work, some basic functions on automata as well as functions on the sets of words have to be defined. We assume that below given routines are available throughout the whole book. Please refer to Appendixes A, B, and C in order to familiarize with the Python programming language, its packages relevant to automata, grammars, and regexes, and some combinatorial optimization tools. Please notice also that we follow the docstring convention. A docstring is a string literal that occurs as the first statement in a function (module, class, or method definition). Such string literals act as documentation.

```python
from FAdo.fa import *

def alphabet(S):
  """Finds all letters in S
  Input: a set of strings: S
  Output: the alphabet of S"""
  result = set()
  for s in S:
    for a in s:
      result.add(a)
  return result

def prefixes(S):
  """Finds all prefixes in S
  Input: a set of strings: S
  Output: the set of all prefixes of S"""
  result = set()
  for s in S:
    for i in xrange(len(s) + 1):
      result.add(s[:i])
  return result

def suffixes(S):
  """Finds all suffixes in S
  Input: a set of strings: S
  Output: the set of all suffixes of S"""
```

© Springer International Publishing AG 2017

W. Wieczorek, *Grammatical Inference*, Studies in Computational Intelligence 673,

DOI 10.1007/978-3-319-46801-3_2

```
    result = set()
    for s in S:
      for i in xrange(len(s) + 1):
        result.add(s[i:])
    return result

def catenate(A, B):
  """Determine the concatenation of two sets of words
  Input: two sets (or lists) of strings: A, B
  Output: the set AB"""
  return set(a+b for a in A for b in B)

def ql(S):
    """Returns the list of S in quasi-lexicographic order
    Input: collection of strings
    Output: a sorted list"""
    return sorted(S, key = lambda x: (len(x), x))

def buildPTA(S):
  """Build a prefix tree acceptor from examples
  Input: the set of strings, S
  Output: a DFA representing PTA"""
  A = DFA()
  q = dict()
  for u in prefixes(S):
    q[u] = A.addState(u)
  for w in iter(q):
    u, a = w[:-1], w[-1:]
    if a != '':
      A.addTransition(q[u], a, q[w])
    if w in S:
      A.addFinal(q[w])
  A.setInitial(q[''])
  return A

def merge(q1, q2, A):
  """Join two states, i.e., q2 is absorbed by q1
  Input: q1, q2 state indexes and an NFA A
  Output: the NFA A updated"""
  n = len(A.States)
  for q in xrange(n):
    if q in A.delta:
      for a in A.delta[q]:
        if q2 in A.delta[q][a]: A.addTransition(q, a, q1)
    if q2 in A.delta:
      for a in A.delta[q2]:
        if q in A.delta[q2][a]: A.addTransition(q1, a, q)
  if q2 in A.Initial: A.addInitial(q1)
  if q2 in A.Final: A.addFinal(q1)
  A.deleteStates([q2])
  return A

def accepts(w, q, A):
  """Verify if in an NFA A, a state q recognizes given word
  Input: a string w, a state index (int) q, and an NFA A
  Output: yes or no as Boolean value"""
  ilist = A.epsilonClosure(q)
  for c in w:
```

Fig. 2.1 A PTA accepting aa, aba, and bba

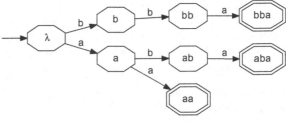

Fig. 2.2 An NFA after merging a and ab in PTA

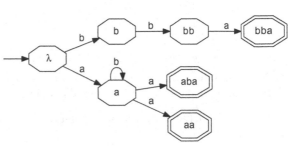

```
    ilist = A.evalSymbol(ilist, c)
    if not ilist:
        return False
return not A.Final.isdisjoint(ilist)
```

There are two fundamental functions that are present in every state merging algorithms given in this book: `buildPTA` for constructing a prefix tree acceptor and `merge` which performs the merging operation.

Definition 2.1 A *prefix tree acceptor* (PTA) is a tree-like DFA built from the learning examples S by taking all the prefixes in the examples as states and constructing the smallest DFA A which is a tree which holds $L(A) = S$. The initial state is a root and all remaining states q have exactly one ingoing edge, i.e., $|\{q' : q \in \delta(q', a)\}| = 1$.

An exemplary PTA is depicted in Fig. 2.1.

The merging operation takes two states from an NFA and joins them into a single state. As we can see from the definition of the `merge` function, the new state (which inherits a label after `q1`) shares the properties, as well as ingoing and outgoing arcs of both states that have been merged. Consider for instance automaton from Fig. 2.1. If states `a` and `ab` are merged, resulting automaton is as in Fig. 2.2.

It should be noted that after this operation the PTA lost the determinism property and—what is more attractive—the new automaton represents an infinite language.

2.2 Evidence Driven State Merging

The idea behind this algorithm is fairly straightforward. Given a sample, we start from building a PTA based on examples, then iteratively select two states and do merging unless compatibility is broken. A heuristic for choosing the pair of states to merge, can be realized in many ways. We propose the following procedure. A score is given to each state pair, and the state pair with the best score is chosen. In order to explain the score in the simplest way (and for further investigations in the present book), we ought to define the right and the left languages of a state q.

Definition 2.2 For the state $q \in Q$ of an NFA $A = (Q, \Sigma, \delta, s, F)$ we consider the two languages:

$$\overrightarrow{L}(q) = \{w \in \Sigma^*: \delta(q, w) \cap F \neq \emptyset\}, \quad \overleftarrow{L}(q) = \{w \in \Sigma^*: q \in \delta(s, w)\}.$$

Thus, the right language of a state q, $\overrightarrow{L}(q)$, is the set of all words spelled out on paths from q to a final state, whereas the left language of a state q, $\overleftarrow{L}(q)$, is the set of all words spelled out on paths from the initial state s to q.

Now we can define the score of two states $q, r \in Q$ for an NFA A and the set U of the suffixes of S_+:

$$\text{score}(q, r) = |U \cap \overrightarrow{L}(q) \cap \overrightarrow{L}(r)|.$$

Finally, we have got the following form of the EDSM algorithm:

```python
def makeCandidateStatesList(U, A):
    """Build the sorted list of pairs of states to merge
    Input: a set of suffixes, U, and an NFA, A
    Output: a list of pairs of states, first most promising"""
    n = len(A.States)
    score = dict()
    langs = []
    pairs = []
    for i in xrange(n):
        langs.append(set(u for u in U if accepts(u, i, A)))
    for i in xrange(n-1):
        for j in xrange(i+1, n):
            score[i, j] = len(langs[i] & langs[j])
            pairs.append((i, j))
    pairs.sort(key = lambda x: -score[x])
    return pairs


def synthesize(S_plus, S_minus):
    """Infers an NFA consistent with the sample
    Input: the sets of examples and counter-examples
    Output: an NFA"""
    A = buildPTA(S_plus).toNFA()
    U = suffixes(S_plus)
    joined = True
    while joined:
        pairs = makeCandidateStatesList(U, A)
        joined = False
        for (p, q) in pairs:
            B = A.dup()
            merge(p, q, B)
            if not any(B.evalWordP(w) for w in S_minus):
```

```
        A = B
        joined = True
        break
return A
```

2.3 Gold's Idea

The central structure of the present algorithm is a table, which during the run is expanded vertically. Its columns are indexed by all suffixes (called EXP) of a sample $\Sigma^* \supset S = (S_+, S_-)$ and its rows are indexed by a prefixed closed set starting from the set $\{\lambda\} \cup \Sigma$ (as usual Σ denotes an alphabet). The rows of the table correspond to the states of the final deterministic automaton A. The indexes (words) of the rows are divided into two sets: RED and BLUE. The RED indexes correspond to states that have been analyzed and which will not be revisited. The BLUE indexes are the candidate states: they have not been analyzed yet and it should be from this set that a state is drawn in order to consider merging it with a RED state. There are three types of entries in the table, which we will call observation table (OT). $OT[u, e] = 1$ if $ue \in L(A)$; $OT[u, e] = 0$ if $ue \notin L(A)$; and $OT[u, e] = *$ otherwise (not known). The sign $*$ is called a hole and corresponds to a missing observation.

Definition 2.3 Rows indexed by u and v are *obviously different* (OD) for OT if there exists such $e \in EXP$ that $OT[u, e]$, $OT[v, e] \in \{0, 1\}$ and $OT[u, e] \neq OT[v, e]$.

Definition 2.4 An observation table OT is *complete* (or has no holes) if for every $u \in$ RED $\cup$ BLUE and for every $e \in$ EXP, $OT[u, e] \in \{0, 1\}$.

Definition 2.5 A table OT is *closed* if for every $u \in$ BLUE there exists $s \in$ RED such that for every $e \in$ EXP $OT[u, e] = OT[s, e]$.

The algorithm is divided into four phases. We will illustrate its run by an example. Let $S = (\{\lambda, ab, abab\}, \{a, b, aa, ba, bb, aab, bab, bbb\})$ be a sample for which we want to find a consistent DFA.

Building a table from the data

```
def buildTable(S_plus, S_minus):
  """Builds an initial observation table
  Input: a sample
  Output: OT as dictionary and sets: Red, Blue, EXP"""
  OT = dict()
  EXP = suffixes(S_plus | S_minus)
  Red = {''}
  Blue = alphabet(S_plus | S_minus)
  for p in Red | Blue:
    for e in EXP:
      if p+e in S_plus:
        OT[p, e] = 1
      else:
        OT[p, e] = 0 if p+e in S_minus else '*'
  return (Red, Blue, EXP, OT)
```

This phase is easy. For a sample S we get the following table:

	' '	a	b	aa	ab	ba	bb	aab	bab	bbb	abab
' '	1	0	0	0	1	0	0	0	0	0	1
a	0	0	1	*	0	*	*	*	1	*	*
b	0	0	0	*	0	*	0	*	*	*	*

Throughout the rest of the example run, RED rows occupy the upper part, while BLUE rows occupy the lower part of the table.

Updating the table

This phase is performed through the while loop that we can see in the `synthesize` function given at the end. The aim of this phase is to bring about the table to be closed. To this end, every BLUE word (index) b that is obviously different from all RED words becomes RED and rows indexed by ba, for $a \in \Sigma$, are added to the BLUE part of the table. In the example, this operation has been repeated two times (for a and then for b):

	' '	a	b	aa	ab	ba	bb	aab	bab	bbb	abab
' '	1	0	0	0	1	0	0	0	0	0	1
a	0	0	1	*	0	*	*	*	1	*	*
b	0	0	0	*	0	*	0	*	*	*	*
aa	0	*	0	*	*	*	*	*	*	*	*
ab	1	*	*	*	1	*	*	*	*	*	*

	' '	a	b	aa	ab	ba	bb	aab	bab	bbb	abab
' '	1	0	0	0	1	0	0	0	0	0	1
a	0	0	1	*	0	*	*	*	1	*	*
b	0	0	0	*	0	*	0	*	*	*	*
aa	0	*	0	*	*	*	*	*	*	*	*
ab	1	*	*	*	1	*	*	*	*	*	*
ba	0	*	0	*	*	*	*	*	*	*	*
bb	0	*	0	*	*	*	*	*	*	*	*

Filling in the holes

Now, the table is closed. The next phase is in order to make the table complete.

```
def fillHoles(Red, Blue, EXP, OT):
  """Tries to fill in holes in OT
  Input: rows (Red + Blue), columns (EXP), and table (OT)
  Output: true if success or false if fail"""
  for b in ql(Blue):
    found = False
    for r in ql(Red):
      if not any(OT[r, e] == 0 and OT[b, e] == 1 \
        or OT[r, e] == 1 and OT[b, e] == 0 for e in EXP):
        found = True
        for e in EXP:
          if OT[b, e] != '*':
            OT[r, e] = OT[b, e]
    if not found:
      return False
  for r in Red:
    for e in EXP:          .
      if OT[r, e] == '*':
        OT[r, e] = 1
  for b in ql(Blue):
    found = False
    for r in ql(Red):
      if not any(OT[r, e] == 0 and OT[b, e] == 1 \
        or OT[r, e] == 1 and OT[b, e] == 0 for e in EXP):
        found = True
        for e in EXP:
          if OT[b, e] == '*':
            OT[b, e] = OT[r, e]
    if not found:
      return False
  return True
```

This routine, first fills the rows corresponding to the RED states by using the information included in the BLUE rows which do not cause any conflict. In our example, the table has not been changed. Then all the holes in the RED rows are filled by 1s:

	' '	a	b	aa	ab	ba	bb	aab	bab	bbb	abab
' '	1	0	0	0	1	0	0	0	0	0	1
a	0	0	1	1	0	1	1	1	1	1	1
b	0	0	0	1	0	1	0	1	1	1	1
aa	0	*	0	*	*	*	*	*	*	*	*
ab	1	*	*	*	1	*	*	*	*	*	*
ba	0	*	0	*	*	*	*	*	*	*	*
bb	0	*	0	*	*	*	*	*	*	*	*

Finally, the routine again visits BLUE rows, tries to find a compatible RED row and copies the corresponding entries. This results in the following table:

	''	a	b	aa	ab	ba	bb	aab	bab	bbb	abab
''	1	0	0	0	1	0	0	0	0	0	1
a	0	0	1	1	0	1	1	1	1	1	1
b	0	0	0	1	0	1	0	1	1	1	1
aa	0	0	0	1	0	1	0	1	1	1	1
ab	1	0	0	0	1	0	0	0	0	0	1
ba	0	0	0	1	0	1	0	1	1	1	1
bb	0	0	0	1	0	1	0	1	1	1	1

Building a DFA from a complete and closed table

```python
def buildAutomaton(Red, Blue, EXP, OT):
  """Builds a DFA from closed and complete observation table
  Input: rows (Red + Blue), columns (EXP), and table (OT)
  Output: a DFA"""
  A = DFA()
  A.setSigma(alphabet(Red | Blue | EXP))
  q = dict()
  for r in Red:
    q[r] = A.addState(r)
  for w in Red | Blue:
    for e in EXP:
      if w+e in Red and OT[w, e] == 1:
        A.addFinal(q[w+e])
  for w in iter(q):
    for u in iter(q):
      for a in A.Sigma:
        if all(OT[u, e] == OT[w+a, e] for e in EXP):
          A.addTransition(q[w], a, q[u])
  A.setInitial(q[''])
  return A

def OD(u, v, EXP, OT):
  """Checks if rows u and v obviously different for OT
  Input: two rows (prefixes), columns, and table
  Output: boolean answer"""
  return any(OT[u, e] in {0, 1} and OT[v, e] in {0, 1} \
    and OT[u, e] != OT[v, e] for e in EXP)

def synthesize(S_plus, S_minus):
  """Infers a DFA consistent with the sample
  Input: the sets of examples and counter-examples
  Output: a DFA"""
  (Red, Blue, EXP, OT) = buildTable(S_plus, S_minus)
  Sigma = alphabet(S_plus | S_minus)
  x = ql(b for b in Blue if all(OD(b, r, EXP, OT) for r in Red))
  while x:
    Red.add(x[0])
    Blue.discard(x[0])
    Blue.update(catenate({x[0]}, Sigma))
    for u in Blue:
      for e in EXP:
        if u+e in S_plus:
          OT[u, e] = 1
```

Fig. 2.3 The result

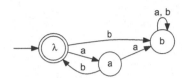

```
      else:
          OT[u, e] = 0 if u+e in S_minus else '*'
    x = ql(b for b in Blue if all(OD(b, r, EXP, OT) for r in Red))
  if not fillHoles(Red, Blue, EXP, OT):
    return buildPTA(S_plus)
  else:
    A = buildAutomaton(Red, Blue, EXP, OT)
    if all(A.evalWordP(w) for w in S_plus) \
      and not any(A.evalWordP(w) for w in S_minus):
      return A
    else:
      return buildPTA(S_plus)
```

In the last phase, the table is transformed into a DFA. RED words constitute the set of states, whereas the transition function is defined using the entries. The algorithm returns a DFA depicted in Fig. 2.3:

```
>>> A = synthesize({"", "ab", "abab"}, \
...    {"a", "b", "aa", "ba", "bb", "aab", "bab", "bbb"})
>>> print A.dotFormat()
digraph finite_state_machine {
  node [shape = doublecircle]; "";
  node [shape = circle]; "a";
  node [shape = circle]; "b";
  ""  -> "a" [label = "a"];
  ""  -> "b" [label = "b"];
  "a" ->  "" [label = "b"];
  "a" -> "b" [label = "a"];
  "b" -> "b" [label = "a, b"];
}
```

2.4 Grammatical Inference with MDL Principle

The minimum description length (MDL) principle is a rule of thumb in which the best hypothesis for a given set of data is the one that leads to the best compression of the data. Generally, searching for small acceptor compatible with examples and counter-examples is a good idea in grammatical inference. MDL principle is the development of this line of reasoning in case of the absence of counter-examples.

Fig. 2.4 Two automata
compatible with a sample S

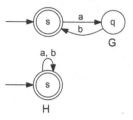

2.4.1 The Motivation and Appropriate Measures

Suppose that we are given the sample $S = \{\lambda, \mathtt{ab}, \mathtt{abab}, \mathtt{ababab}\}$. Let us try
to answer the question what makes an automaton G better than H (see Fig. 2.4)
in describing the language represented by S. The idea is to measure an automaton
along with the size of encoding all words of the sample. In this way, not always is
the smallest automaton recognized as the most promising. The process of parsing is
also at stake here.

Let $A = (Q, \Sigma, \delta, s, F)$ be an NFA all of whose non-final states have outgoing
edges. The number of bits required to encode the path followed to parse a word w
can be assessed by the below given function ch. We associate with each state $q \in Q$
the value $t_q = \sum_{a \in \Sigma} |\delta(q, a)|$ if $q \notin F$. If $q \in F$ then $t_q = 1 + \sum_{a \in \Sigma} |\delta(q, a)|$,
since one more choice is available. We are now in a position to define $\mathrm{ch}(q, w)$. For
the empty word we have $\mathrm{ch}(q, \lambda) = \log(t_q)$ if $q \in F$; otherwise $\mathrm{ch}(q, \lambda) = \infty$.
For $w = au$ ($a \in \Sigma$, $u \in \Sigma^*$) ch depends on the recursive definition: $\mathrm{ch}(q, w) =
\log(t_q) + \min_{r \in \delta(q,a)} \mathrm{ch}(r, u)$ and $\mathrm{ch}(q, w) = \infty$ if $\delta(q, a) = \emptyset$.

We can now, given a sample S and an NFA A, measure the score sc of A:

$$\mathrm{sc}(A, S) = |Q| + \|\delta\|(2 \log |Q| + \log |\Sigma|) + \sum_{w \in S} \mathrm{ch}(s, w),$$

where $\|\delta\|$ is the number of transitions of A.

Coming back to automata G, H and a sample S, the exact computations of the
scores can be found below. Notice that $t_s = 2$ and $t_q = 1$ for G, while $t_s = 3$ for
H. Thus, we have $\mathrm{sc}(G, S) = 2 + 2(2 \log(2) + \log(2)) + 10 \log(2) = 18.0$ and
$\mathrm{sc}(H, S) = 1 + 2(2 \log(1) + \log(2)) + 16 \log(3) = 28.36$. Therefore, we need more
space to encode both the automaton and the data in case of H and S than in case of
G and S.

2.4.2 The Proposed Algorithm

The idea is as follows. We start from the PTA and iteratively merge a pair of states as
long as this operation decreases the score. The order in which states are merged can
be scheduled in many ways. We base on the quasi-lexicographic order of the labels of

states. After building the PTA, every state in an automaton has a label corresponding
to the path from the initial state to that state.

```python
from math import log
from FAdo.fa import *


def sc(A, S):
  """Measures the score of an NFA A and words S
  Input: an automaton A and the set of words S
  Output: a float"""

  @memoize
  def ch(i, w):
    """Calculates the size of encoding of the path
    followed to parse word w from the ith state in A
    Input: state's index, word
    Output: a float"""
    if w == '':
      return log(t[i], 2) if i in A.Final else float('inf')
    else:
      if i in A.delta and w[0] in A.delta[i]:
        return log(t[i], 2) + min(ch(j, w[1:]) \
          for j in A.delta[i][w[0]])
      else:
        return float('inf')

  s = list(A.Initial)[0]
  t = dict()
  for i in xrange(len(A.States)):
    t[i] = 1 if i in A.Final else 0
    if i in A.delta:
      t[i] += sum(map(len, A.delta[i].itervalues()))
  return len(A.States) + sum(ch(s, w) for w in S) \
    + A.countTransitions()*(2*log(len(A.States), 2) \
    + log(len(A.Sigma), 2))


def synthesize(S):
  """Finds a consistent NFA by means of the MDL principle
  Input: set of positive words
  Output: an NFA"""
  A = buildPTA(S).toNFA()
  Red = {''}
  Blue = set(A.States)
  Blue.remove('')
  current_score = sc(A, S)
  while Blue:
    b = ql(Blue)[0]
    Blue.remove(b)
    for r in ql(Red):
      M = A.dup()
      merge(M.States.index(r), M.States.index(b), M)
      new_score = sc(M, S)
      if new_score < current_score:
        A = M
        current_score = new_score
        break
    if b in A.States:
      Red.add(b)
  return A
```

2.5 Bibliographical Background

The evidence driven state merging algorithm given in Sect. 2.2 is based in part on concepts published in Coste and Fredouille (2003). Gold's algorithm (Gold 1978) was the first GI algorithm with convergence properties. We have presented the algorithm according to its description in de la Higuera (2010). The MDL principle in the context of deterministic automata was described in de la Higuera (2010). We have adapted it to a non-deterministic output in Sect. 2.4. More thorough theoretical investigations in this regard were carried by Adriaans and Jacobs (2006) for DFAs, and by Petasis et al. (2004) for CFGs.

It is also worth to note three state of the art tools for heuristic state-merging DFA induction: the Trakhtenbrot-Barzdin state merging algorithm (denoted Traxbar) adapted by Lang (1992), Rodney Price's Abbadingo winning idea of evidence-driven state merging (Blue-fringe) described by Lang et al. (1998), and Rlb state merging algorithm (Lang 1997).

Trakhtenbrot and Barzdin (1973) described an algorithm for constructing the smallest DFA consistent with a complete labeled training set. The input to the algorithm is the PTA. This tree is squeezed into a smaller graph by merging all pairs of states that represent compatible mappings from word suffixes to labels. This algorithm for completely labeled trees was generalized by Lang (1992) to produce a (not necessarily minimum) machine consistent with a sparsely labeled tree.[1]

The second algorithm that starts with the PTA and folds it up into a compact hypothesis by merging pairs of states is Blue-fringe. This program grows a connected set of red nodes that are known to be unique states, surrounded by a fringe of blue nodes that will either be merged with red nodes or be promoted to red status. Merges only occur between red nodes and blue nodes. Blue nodes are known to be the roots of trees, which greatly simplifies the code for correct merging. The only drawback of this approach is that the pool of possible merges is small, so occasionally the program has to do a low scoring merge.

The idea that lies behind the third algorithm, Rlb, is as follows. It dispenses with the red-blue restriction and is able to do merges in any order. However, to have a practical run time, only merges between nodes that lie within a distance 'window' of the root on a breadth-first traversal of the hypothesis graph are considered. This introduction of a new parameter is a drawback to this program, as is the fact that its run time scales very badly with training string length. However, on suitable problems, it works better than the Blue-fringe algorithm. In Lang (1997) one can find the detailed description of heuristics for evaluating and performing merges.

---

[1] The reader can use implementations from the archive http://abbadingo.cs.nuim.ie/dfa-algorithms. tar.gz for the Traxbar and for the two remaining state-merging algorithms.

References

Adriaans PW, Jacobs C (2006) Using MDL for grammar induction. In: Proceedings of grammatical inference: algorithms and applications, 8th international colloquium, ICGI 2006. Tokyo, Japan, 20–22 Sept 2006, pp 293–306

Coste F, Fredouille D (2003) Unambiguous automata inference by means of state-merging methods. In: Proceedings of machine learning: ECML 2003, 14th European conference on machine learning. Cavtat-Dubrovnik, Croatia, 22–26 Sept 2003, pp 60–71

de la Higuera C (2010) Grammatical inference: learning automata and grammars. Cambridge University Press, New York, NY, USA

Gold EM (1978) Complexity of automaton identification from given data. Inf Control 37:302–320

Lang KJ (1992) Random DFA's can be approximately learned from sparse uniform examples. In: Proceedings of the fifth annual workshop on computational learning theory. ACM, pp 45–52

Lang KJ (1997) Merge order count. Technical report, NECI

Lang KJ, Pearlmutter BA, Price RA (1998) Results of the abbadingo one DFA learning competition and a new evidence-driven state merging algorithm. In: Proceedings of the 4th international colloquium on grammatical inference. Springer, pp 1–12

Petasis G, Paliouras G, Karkaletsis V, Halatsis C, Spyropoulos CD (2004) E-GRIDS: computationally efficient grammatical inference from positive examples. Grammars 7:69–110

Trakhtenbrot B, Barzdin Y (1973) Finite automata: behavior and synthesis. North-Holland Publishing Company

Chapter 3
Partition-Based Algorithms

3.1 Preliminaries

The common operation performed in partition-based algorithms is the merging of groups of states (or non-terminals if we deal with CFGs). These groups come from a partition. Naturally, we start from the definition of a partition and the definition of the NFA (and CFG) induced by a partition.

Definition 3.1 A family of sets π is a *partition* of X if and only if all of the following conditions hold: (i) π does not contain the empty set, (ii) the union of the sets in π is equal to X, and (iii) π is pairwise disjoint. The sets in π are called the *blocks*, parts or cells of the partition. If π is a partition of X, then for any element $x \in X$ there is a unique block containing x, which we denote $K(x, \pi)$.

Definition 3.2 Let $A = (Q, \Sigma, \delta, s, F)$ be an NFA and π be a partition of the set Q of states. We define the NFA $A/\pi = (Q', \Sigma, \delta', s', F')$ to be *induced by π from A* as follows:
$$Q' = \pi,$$
$$K(r, \pi) \in \delta'(K(q, \pi), a) \quad \text{if } r \in \delta(q, a),$$
$$s' = K(s, \pi),$$
$$F' = \{K(q, \pi): q \in F\}.$$

The definition of a context-free grammar imposes no restriction whatsoever on the right side of a production. A *normal form* for context-free grammars is one that, although restricted, is broad enough so that any grammar has an equivalent normal form version. Amongst all normal forms for context-free grammars, the most useful and the most well-known one is the Chomsky normal form (CNF).

Definition 3.3 A CFG $G = (V, \Sigma, P, S)$ is said to be in *Chomsky normal form* if each of its rules is in one of the two possible forms:

1. $A \rightarrow a, \quad a \in \Sigma, A \in V,$
2. $A \rightarrow BC, \quad A, B, C \in V.$

© Springer International Publishing AG 2017
W. Wieczorek, *Grammatical Inference*, Studies in Computational Intelligence 673,
DOI 10.1007/978-3-319-46801-3_3

Definition 3.4 Let $G = (V, \Sigma, P, S)$ be a CFG in CNF and π be a partition of the set V of variables $(A, B, C \in V, a \in \Sigma)$. The CFG $G/\pi = (V', \Sigma, P', S')$ is *induced by π from G* if:

$$V' = \pi,$$
$$P' = \{K(A, \pi) \to K(B, \pi)\, K(C, \pi) \colon (A \to B\,C) \in P\}$$
$$\cup \{K(A, \pi) \to a \colon (A \to a) \in P\},$$
$$S' = K(S, \pi).$$

Definition 3.5 The finite set X of words is said to be *structurally complete* with respect to a finite automaton $A = (Q, \Sigma, \delta, s, F)$ if the two following conditions hold:

1. Every transition of A is used at least once when accepting the words of X.
2. Every element of F is an accepting state of at least one word.

The following theorem forms the basis of a family of the regular inference, which we call partition-based methods.

Theorem 3.1 *Let S_+ be the positive part of a sample of any regular language R and let A be an automaton isomorphic to the minimal DFA accepting R. If S_+ is structurally complete with respect to A, then A may be derived from $PTA(S_+)$ for some partition π.*

The representation of a CFG

In order to keep balance between simplicity and efficiency, we have decided to store a CFG as Python's set of tuples. Each tuple represents one production rule and may be either (A, B, C) or (A, a), where A, B, C are non-negative integers and a is a one-letter string. These tuples correspond to productions $A \to B\,C$ and $A \to a$. In addition to this, the start symbol is always represented by the number 0.

The following class will be helpful in checking whether a word is accepted by such a grammar:

```python
class Parser(object):
    """A parser class for CNF grammars"""

    def __init__(self, productions):
        self.__prods = productions
        self.__cache = dict()

    def __parse(self, w, var):
        """Unger's parsing method
        Input: a word, a non-terminal (an integer)
        Output: true or false"""
        if (w, var) in self.__cache:
            return self.__cache[w, var]
        else:
            n = len(w)
            if n == 1: return (var, w) in self.__prods
            for p in self.__prods:
                if p[0] == var:
```

```
            if len(p) == 3:
              for i in xrange(1, n):
                if self.__parse(w[:i], p[1]) \
                   and self.__parse(w[i:], p[2]):
                  self.__cache[w, var] = True
                  return True
        self.__cache[w, var] = False
        return False

  def accepts(self, word):
    """Membership query
    Input: a string
    Output: true or false"""
    self.__cache.clear()
    return self.__parse(word, 0)

  def grammar(self):
    return self.__prods
```

The implementation of the induction operations

In the first two of presented algorithms, we will deal with NFAs induced by a partition.
The implementation of achieving A/π is as follows:

```
from FAdo.fa import *

def inducedNFA(P, A):
  """Join groups of states for a new automaton
  Input: the partition P (the list of frozensets) and an NFA A
  Output: a new NFA, A/P"""
  B = NFA()
  d = dict()
  K = range(len(A.States))
  for p in P:
    d[p] = j = B.addState(p)
    if p & A.Initial:
      B.addInitial(j)
    if p & A.Final:
      B.addFinal(j)
    for state in p:
      K[state] = j
  for q in A.delta:
    for a in A.delta[q]:
      for r in A.delta[q][a]:
        B.addTransition(K[q], a, K[r])
  return B
```

In the last algorithm CFGs are induced by sorted partitions from a settled CFG G.
In a sorted partition the element 0 is always included in the first set of the partition,
which is essential because the first block (of index 0) will become the new start
symbol. The following function performs G/π:

```
def inducedCFG(P, G):
  """Join groups of variables for a new CFG
  Input: a sorted partition P and productions G in CNF
  Output: a new CFG in Chomsky normal form, G/P"""
  K = dict((v, i) for i in xrange(len(P)) for v in P[i])
  Pprime = set()
```

```
for p in G:
  if len(p) == 3:
    (A, B, C) = p
    Pprime.add((K[A], K[B], K[C]))
  elif isinstance(p[1], str):
    (A, a) = p
    Pprime.add((K[A], a))
return Pprime
```

3.2   The k-tails Method

What distinguishes the k-tails method from other GI methods is the series of automata on output instead of a single one. On input we give only positive words X. Let k be a non-negative integer and $A = (Q, \Sigma, \delta, s, F)$ be an NFA. The equivalence relation defined by $p \equiv q$ if and only if $\overrightarrow{L}(p) \cap \Sigma^{\leq k} = \overrightarrow{L}(q) \cap \Sigma^{\leq k}$ helps in making a partition. As usual, the equivalence class of any element $q \in Q$, denoted by $[q]$, is the set of elements to which q is related through the equivalence relation $\equiv$. The partition of the states of A is defined by $\pi = \{[q] : q \in Q\}$.

At the beginning, the algorithm finds the smallest (in terms of the number of states) DFA[1] M such that $L(M) = X$, and then for $k = 0, 1, \ldots, m - 1$, where m is the length of the longest word in X, generates M/π, one after the other.

```
def synthesize(X):
  """Finds NFAs for k = 0, 1, ... by means of the k-tails method
  Input: the set of strings
  Output: an iterator over NFAs"""
  minanfa = buildPTA(X).minimalHopcroft().toNFA()
  n = len(minanfa.States)
  m = max(len(x) for x in X)
  S = suffixes(X)
  langs = []
  for i in xrange(n):
    langs.append(set(w for w in S if accepts(w, i, minanfa)))
  for k in xrange(m):
    d = dict()
    for i in xrange(n):
      el = frozenset(w for w in langs[i] if len(w) <= k)
      if el in d:
        d[el].add(i)
      else:
        d[el] = {i}
    partition = map(frozenset, d.itervalues())
    yield inducedNFA(partition, minanfa)
```

It is worth stressing that, the larger k, the more particular and extensive hypothesis are yielded.

---

[1]Notice that a DFA is the special kind of an NFA. It is possible to define two different DFAs A and B such that $L(A) = L(B)$. A natural question arises from the ambiguity mentioned above: is there, for a given regular language, only one DFA with the minimum number of states? The answer is affirmative. For each DFA we can find an equivalent DFA that has as few states as any DFA accepting the same language. Moreover, except for our ability to call the states by whatever names we choose, this minimum-state DFA is unique to the language.

3.3 Grammatical Inference by Genetic Search

In the present method we also deal with some initial automaton and with the partitions of its states. This time, however, we start from the prefix tree acceptor T of the positive part of a sample. The second difference is that the optimal partition is determined by a genetic algorithm.

3.3.1 What Are Genetic Algorithms?

A genetic algorithm makes it possible to solve an optimization problem without a tedious phase of constructing a man-made program which solves it. Its work is modeled upon the natural microevolution of organisms. The evolution proceeds in accordance to the principle of survival and reproduction of the fittest. A population in genetic algorithms is a multiset[2] of problem solutions (called chromosomes or individuals) usually represented as sequences of bits or numbers. Such a population evolves as the result of repeatedly executed operations: the selection of best solutions and the creation of new ones out of them. While creating new solutions the operators of recombination such as crossover and mutation are used. The new solutions replace other solutions in the population.

3.3.2 Basic Notions of the Genetic Algorithm for GI

There are four preparatory steps which must be accomplished before a searching process for a program to solve the problem can begin. These are as follows: (a) choice of a way to represent partitions as sequences, (b) defining the fitness function, (c) defining the control parameters, and (d) defining the termination criterion.

The choice how partitions are encoded and a definition of the fitness function determine to a large extent the solution space which will be searched. The control parameters include the population size, the probabilities of crossover and mutation, the choice of a selection scheme, etc.

Partition representation scheme

Let $Q = \{0, 1, \ldots, m - 1\}$ be states' indexes. Each partition π of Q is encoded as a sequence of m integers $(i_0, i_1, \ldots, i_{m-1})$ where $i_j \in 0, 1, \ldots, m - 1, 0 \leq j < m$. The jth integer i_j indicates the block assigned to the state j, i.e., $i_p = i_q$ if and only if $K(p, \pi) = K(q, \pi)$. For example, the chromosome $(3, 2, 3, 0, 3)$ represents the partition $\{\{0, 2, 4\}, \{1\}, \{3\}\}$ of the set $\{0, 1, 2, 3, 4\}$.

---

[2]A multiset (or bag) is a generalization of the concept of a set that, unlike a set, allows multiple instances of the multiset's elements.

In the literature on genetic algorithms, several other representations of partitions are proposed (base upon permutations, for instance), but above mentioned one has the closure property. It means that any recombination operation keeps a chromosome valid. In the final program we will use the default genetic operators of the Pyevolve[3] package: swap mutator and one point crossover. Suppose that we have the following chromosomes:

$$a = (2, 2, 0, 1, 3, 5),$$
$$b = (2, 1, 2, 3, 2, 4).$$

They represent the partitions: $\{\{0, 1\}, \{2\}, \{3\}, \{4\}, \{5\}\}$ and $\{\{0, 2, 4\}, \{1\}, \{3\}, \{5\}\}$. The mutation operation is performed on a single chromosome and relies on selecting at random two indexes in the sequence and then swapping their integers. For example

$$a' = (2, 3, 0, 1, 2, 5)$$

is a swapped a and represents $\{\{0, 4\}, \{1\}, \{2\}, \{3\}, \{5\}\}$. In the crossover operation two chromosomes are involved. A random index is selected. The first part of the first chromosome is connected with the second part of the second chromosome to make the first offspring. The second offspring is build from the first part of the second chromosome and the second part of the first chromosome (the crossover point is noted by the | sign):

$$a' = (2, 2, 0, 1 \mid 2, 4),$$
$$b' = (2, 1, 2, 3 \mid 3, 5).$$

The new chromosomes represent, respectively, the partitions: $\{\{0, 1, 4\}, \{2\}, \{3\}, \{5\}\}$ and $\{\{0, 2\}, \{1\}, \{3, 4\}, \{5\}\}$.

Fitness function

The aim of the fitness function is to provide the basis for competition among individuals of a population. It is important that not only should the correct solutions obtain a high assessment (reward), but also every improvement of an individual should result in increasing of that reward (or decreasing cost).

For the GI problem considered herein, the raw fitness score (i.e., the value returned by fitness function before any scaling that might be required for a specific selection scheme) is calculated as follows. If $L(T/\pi) \cap S_- \neq \emptyset$ then the score gets infinity; otherwise the score gets the number of blocks in π. Of course, the lower raw fitness, the more favorable outcome.

---

[3] See Appendix B for closer information.

3.3.3 Our Implementation

Such control parameters as the population size and the number of iterations should be selected with considering the complexity of a task, the amount of computer memory, and the total time available for computing. Because genetic algorithms are rather time consuming we decided to apply an incremental procedure described in Sect. 1.1.5.

```python
from FAdo.fa import *
from FAdo.common import DFAsymbolUnknown
from itertools import izip, count
from pyevolve import Consts
from pyevolve import G1DList
from pyevolve import GSimpleGA

negPart = []
PTA = NFA()

def incrementalInference(S_plus, S_minus, f_syn, elem):
  """Semi-incremental procedure for GI
  Input: a sample, a synthesizing function, and a membership
         query function
  Output: a hypothesis consistent with S or None"""
  Sp, Sm = ql(S_plus), ql(S_minus)
  n = len(S_plus)
  j = 2
  I = Sp[0:j]
  while j <= n:
    hypothesis = f_syn(I, Sm)
    k = j
    if hypothesis:
      while k < n:
        if elem(Sp[k], hypothesis): k += 1
        else: break
      if k < n: I.append(Sp[k])
    else: break
    j = k+1
  return hypothesis

def decodeToPi(chromosome):
  """Finds an appropriate partition
  Input: a chromosome
  Output: a partition as the list of frozensets"""
  t = dict()
  for (v, i) in izip(chromosome, count(0)):
    if v in t:
      t[v].add(i)
    else:
      t[v] = {i}
  return map(frozenset, t.itervalues())

def accepted(word, automaton):
  """Checks whether a word is accepted by an automaton
  Input: a string and an NFA
  Output: true or false"""
  try:
    return automaton.evalWordP(word)
```

```python
    except DFAsymbolUnknown:
      return False

def eval_func(chromosome):
  """The fitness function
  Input: a chromosome
  Output: a float"""
  global negPart, PTA
  Pi = decodeToPi(chromosome)
  A = inducedNFA(Pi, PTA)
  if any(accepted(w, A) for w in negPart):
    return float('infinity')
  else:
    return len(Pi)

def synthesizeByGA(positives, negatives):
  """Finds an NFA consistent with the input by means of GA
  Input: a sample, S = (positives, negatives)
  Output: an NFA or None"""
  global negPart, PTA
  negPart = negatives
  PTA = buildPTA(positives).toNFA()
  genome = G1DList.G1DList(len(PTA.States))
  genome.setParams(rangemin=0, rangemax=len(PTA.States)-1)
  genome.evaluator.set(eval_func)
  ga = GSimpleGA.GSimpleGA(genome)
  ga.setGenerations(500)
  ga.setMinimax(Consts.minimaxType["minimize"])
  ga.setCrossoverRate(0.9)
  ga.setMutationRate(0.05)
  ga.setPopulationSize(200)
  ga.evolve()
  best_indiv = ga.bestIndividual()
  if best_indiv.getRawScore() == float('infinity'):
    return None
  else:
    Pi = decodeToPi(best_indiv)
    A = inducedNFA(Pi, PTA)
    return A

def synthesize(S_plus, S_minus):
  """Finds an NFA consistent with the input by means of GA
  using incremental procedure
  Input: a sample, S = (S_plus, S_minus), |S_plus| > 1
  Output: an NFA or None"""
  return incrementalInference(S_plus, S_minus, \
    synthesizeByGA, accepted)
```

3.4 CFG Inference Using Tabular Representations

The last method presented in this chapter, unlike the previous ones, outputs context-free grammars. In the method, a hypothesis about the unknown language is efficiently represented by a table-like data structure similar to the parse table used in

Cocke-Younger-Kasami (CYK) algorithm. By employing this representation method, the induction of a CFG from a sample can be reduced to the problem of partitioning the set of non-terminals.

3.4.1 Basic Definitions

Definition 3.6 Given a word $w = a_1 a_2 \cdots a_n$ of length n, the *tabular representation* for w is the triangular table $T(w)$ where each element, denoted $t_{i,j}$, for $1 \leq i \leq n$ and $2 \leq j \leq n - i + 1$, contains the set $\{X_{i,j,1}, \ldots, X_{i,j,j-1}\}$ of $j - 1$ distinct variables. For $j = 1$, $t_{i,1}$ is the singleton set $\{X_{i,1,1}\}$.

Definition 3.7 The *primitive* CFG $G(T(w)) = (V, \Sigma, P, S)$ derived from the tabular representation $T(w)$ is the following grammar, in Chomsky normal form:

$$V = \{X_{i,j,k} : 1 \leq i \leq n, \ 1 \leq j \leq n - i + 1, \ 1 \leq k < j\}$$
$$\cup \{X_{i,1,1} : 1 \leq i \leq n\},$$
$$P = \{X_{i,j,k} \rightarrow X_{i,k,l_1} X_{i+k,j-k,l_2} : 1 \leq i \leq n,$$
$$1 \leq j \leq n - i + 1, \ 1 \leq k < j, \ 1 \leq l_1 < \max\{2, k\}, \ 1 \leq l_2 < \max\{2, j - k\}\}$$
$$\cup \{X_{i,1,1} \rightarrow a_i : 1 \leq i \leq n\}$$
$$\cup \{S \rightarrow X_{1,n,k} : 1 \leq k \leq \max\{1, n - 1\}\}.$$

3.4.2 The Algorithm

The idea of tabular representations

Suppose that we have a word $w = \text{aba}$. All possible parse trees for this word, when a CFG grammar has no unit productions ($A \rightarrow B$), are depicted in Fig. 3.1. The number of such trees is equal to the number of ways to insert $n - 1$ pairs of parentheses in a word of n letters. E.g., for $n = 3$ there are 2 ways: ((ab)c) or (a(bc)); for $n = 4$ there are 5 ways: ((ab)(cd)), (((ab)c)d), ((a(bc))d), (a((bc)d)), (a(b(cd))). In general, there are $C(n - 1)$ such ways, where $C(k) = \binom{2k}{k}/(k + 1)$ is the kth Catalan number. Thus, the number of all possible grammatical structures for a word w of length n becomes exponential in n. Fortunately, the tabular representation for w and the primitive grammar are of size polynomial in n. All parse trees can be derived[4] from the start symbol of the primitive grammar. For $w = \text{aba}$, $G(T(w))$ (see Fig. 3.2 for $T(w)$) is of the form:

$$S \rightarrow X_{1,3,1} \mid X_{1,3,2} \quad X_{1,3,1} \rightarrow X_{1,1,1} X_{2,2,1} \quad X_{1,3,2} \rightarrow X_{1,2,1} X_{3,1,1}$$
$$X_{1,2,1} \rightarrow X_{1,1,1} X_{2,1,1} \quad X_{2,2,1} \rightarrow X_{2,1,1} x_{3,1,1} \quad X_{1,1,1} \rightarrow a$$
$$X_{2,1,1} \rightarrow b \quad\quad\quad X_{3,1,1} \rightarrow a$$

---

[4]Of course we derive words, not trees, but the order of using productions during the derivation can be depicted as a tree.

Fig. 3.1 All possible
grammatical structures for
the word aba

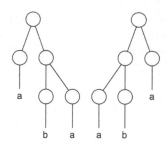

Fig. 3.2 The tabular
representation $T(\text{aba})$

3	$X_{1,3,1}, X_{1,3,2}$		
2	$X_{1,2,1}$	$X_{2,2,1}$	
$j = 1$	$X_{1,1,1}$	$X_{2,1,1}$	$X_{3,1,1}$
	$i = 1$	2	3
	a	b	a

The formulation of the learning algorithm

We assume that a finite set U contains positive, U_+, and negative, U_-, words. In the algorithm, the variables of $G(T(w))$ for each $w \in U_+$ before taking the union of CFGs have to be renamed, so as to avoid the reduplication of symbols (except for the start symbol, S). For a word w, let $G^w(T(w))$ denote the CFG $G(T(w))$ in which every variable X in $G(T(w))$ is renamed to X^w.

The learning scheme is as follows:

1. Construct the tabular representation $T(w)$ for each positive word $w \in U_+$;
2. Derive the primitive CFG $G(T(w))$ and $G^w(T(w))$ for each $w \in U_+$;
3. Take the union of those primitive CFGs, that is,

$$G(T(U_+)) = \bigcup_{w \in U_+} G^w(T(w));$$

4. Find one of the smallest partitions, π, such that $G(T(U_+))/\pi$ is consistent with the positive words U_+ and the negative words U_-;
5. Output the resulting CFG $G(T(U_+))/\pi$.

Fitness function

We define the fitness function f from chromosomes to non-negative real numbers. Let p be a chromosome and π_p be the partition represented by p for the set of non-terminals in $G(T(U_+))$. Let U_+ be the given positive words and U_- be the given negative words. The fitness function f is defined as follows:

$$f_1(p) = \frac{|\{w \in U_+ : w \in G(T(U_+))/\pi_p\}|}{|U_+|},$$

$$f_2(p) = \frac{|p|}{|\pi_p|},$$

$$f(p) = \begin{cases} 0 & \text{if some } w \in U_- \text{ or no } w \in U_+ \text{ is generated} \\ & \text{by } G(T(U_+))/\pi_p, \\ f_1(p) + f_2(p) & \text{otherwise.} \end{cases}$$

3.4.3 Our Implementation

Because partitioning is a computationally intractable problem, we have again decided to take advantage of the genetic algorithm described in the previous section.

```python
from pyevolve import G1DList
from pyevolve import GSimpleGA

class Idx(object):
  """A class for simple indexing
  Declaration: i = Idx(0)
  Usage: n = i+1 or fun(i+1)"""
  def __init__(self, start):
    self.__counter = start
  def __add__(self, step):
    self.__counter += step
    return self.__counter

def tr(w, idx):
  """Returns non-terminals of tabular representation for w
  Input: a string
  Output: the dictionary of non-terminals"""
  n = len(w)
  X = dict()
  for i in xrange(1, n+1):
    X[i, 1, 1] = idx+1
    for j in xrange(2, n-i+2):
      for k in xrange(1, j):
        X[i, j, k] = idx+1
  return X

def primitiveCFG(w, X):
  """Constructs the productions of a primitive CFG
  Input: a string w and nonterminals from a tabular
         representation for w
  Output: the set of productions"""
  n = len(w)
  P = {(0, X[1, n, k]) for k in xrange(1, max(2, n))}
  for i in xrange(1, n+1):
    P.add((X[i, 1, 1], w[i-1]))
    for j in xrange(1, n-i+2):
      for k in xrange(1, j):
        for l1 in xrange(1, max(2, k)):
          for l2 in xrange(1, max(2, j-k)):
            P.add((X[i, j, k], X[i, k, l1], X[i+k, j-k, l2]))
  return P
```

```
def decodeToPi(chromosome):
  """Finds an appropriate partition (a block with 0 first)
  Input: a chromosome
  Output: a sorted partition as the list of frozensets"""
  t = dict()
  n = len(chromosome)
  for (v, i) in zip(chromosome, range(n)):
    if v in t:
      t[v].add(i)
    else:
      t[v] = {i}
  return sorted(map(frozenset, t.itervalues()), key=min)

posPart = set()
negPart = set()
GTU = set()   # G(T(U_+))

def eval_func(chromosome):
  """The fitness function
  Input: a chromosome
  Output: a float"""
  global posPart, negPart, GTU
  Pi = decodeToPi(chromosome)
  G = inducedCFG(Pi, GTU)
  parser = Parser(G)
  if any(parser.accepts(w) for w in negPart):
    return 0.0
  f1 = 0
  for w in posPart:
    if parser.accepts(w):
      f1 += 1
  if f1 == 0: return 0.0
  f1 /= float(len(posPart))
  f2 = float(len(chromosome)) / len(Pi)
  return f1 + f2

def synthesize(Uplus, Uminus):
  """Finds a CFG consistent with the input by means of GA
  Input: a sample, Sample = (Uplus, Uminus) without lambda
  Output: the parser of an inferred CFG or None"""
  global posPart, negPart, GTU
  posPart, negPart = Uplus, Uminus
  idx = Idx(0)
  GTU = set()
  for w in Uplus:
    GTU.update(primitiveCFG(w, tr(w, idx)))
  nvars = idx+1
  genome = G1DList.G1DList(nvars)
  genome.setParams(rangemin = 0, rangemax = nvars-1)
  genome.evaluator.set(eval_func)
  ga = GSimpleGA.GSimpleGA(genome)
  ga.setGenerations(400)
  ga.setCrossoverRate(0.9)
  ga.setMutationRate(0.05)
  ga.setPopulationSize(1000)
  ga.evolve(freq_stats=50)   # will show the statistics
  # every 50th generation (the parameter may be omitted)
```

```
best_indiv = ga.bestIndividual()
Pi = decodeToPi(best_indiv)
G = inducedCFG(Pi, GTU)
parser = Parser(G)
if any(parser.accepts(w) for w in Uminus) \
  or not all(parser.accepts(w) for w in Uplus):
  return None
return parser
```

3.5 Bibliographical Background

The description of grammar's normal forms and their equivalence was presented in Hopcroft et al. (2001) and—especially for the generalization of Greibach normal form—in Wood (1970). Unger's parsing method has been implemented according to Grune and Jacobs (2008), a book that provides us with an access to the full field of parsing techniques. Theorem 3.1 was formulated by Dupont (1994) based on lemmas and theorems proved by Angluin (1982).

The k-tails method, which was described and implemented in Sect. 3.2, bases on a specification given by Miclet (1990). The original idea comes from Biermann and Feldman (1972) and also was later developed by Miclet (1980).

Inductive synthesis of NFAs with the help of genetic algorithms has been presented according to Dupont (1994). For other approaches to encoding partitions the reader is referred to Michalewicz (1996) with a thorough bibliographical research on this topic.

Sakakibara (2005) is the author of the CFG inference method that uses tabular representations. In the work, he also showed that if U_+ is a sufficient representation of an unknown language $L(G_*)$ then there exists a partition π such that $L(G(T(U_+))/\pi) = L(G_*)$. Some improvements on the genetic algorithm solving the partitioning problem—such as generating initial population, adding specialized operators, modifying fitness function—in the context of this algorithm, were made by Unold and Jaworski (2010).

References

Angluin D (1982) Inference of reversible languages. J ACM 29(3):741–765

Biermann AW, Feldman JA (1972) On the synthesis of finite-state machines from samples of their behavior. IEEE Trans Comput 21(6):592–597

Dupont P (1994) Regular grammatical inference from positive and negative samples by genetic search: the GIG method. In: Proceedings of the 2nd international colloquium on grammatical inference, ICGI '94. Springer, Lecture notes in artificial intelligence, vol 862, pp 236–245

Grune D, Jacobs CJ (2008) Parsing techniques: a practical guide, 2nd edn. Springer

Hopcroft JE, Motwani R, Ullman JD (2001) Introduction to automata theory, languages, and computation, 2nd edn. Addison-Wesley

Michalewicz Z (1996) Genetic algorithms + data structures = evolution programs, 3rd edn. Springer

Miclet L (1980) Regular inference with a tail-clustering method. IEEE Trans Syst Man Cybern 10:737–743

Miclet L (1990) Grammatical inference. In: Bunke H, Sanfeliu A (eds) Syntactic and structural pattern recognition: theory and applications, series in computer science, vol 7. World Scientific, chap 9, pp 237–290

Sakakibara Y (2005) Learning context-free grammars using tabular representations. Pattern Recogn 38(9):1372–1383

Unold O, Jaworski M (2010) Learning context-free grammar using improved tabular representation. Appl Soft Comput 10(1):44–52

Wood D (1970) A generalised normal form theorem for context-free grammars. Comput J 13(3):272–277

Chapter 4
Substring-Based Algorithms

4.1 Error-Correcting Grammatical Inference

The algorithm to be presented below, for given $S_+ \subset \Sigma^+$, builds up an NFA $A = (Q, \Sigma, \delta, s, \{f\})$, such that $S_+ \subseteq L(A)$ and which is acyclic (in that $q \notin \delta(q, w)$ for any $q \in Q$ and $w \in \Sigma^+$). The automaton always has the following *ingoing arcs property*: for every $p, q, r \in Q$ and $a, x \in \Sigma$, if $r \neq f$ and $r \in \delta(p, a)$ and $r \in \delta(q, x)$ then $x = a$, i.e., the same symbol is associated with all the edges that end in the same state in the transition diagram of A (except for a final state f). This property is not crucial, yet convenient for the implementation of the described algorithm.

4.1.1 The GI Algorithm

The essential elements of the present algorithm are the three error-correcting rules: insertion, substitution, and deletion. First, we will make the necessary definitions. Let A be an NFA as described above. *Insertion* (also *substitution*) of $x \in \Sigma$ after a state $q \in Q$ ($q \neq f$) makes a new automaton $A' = (Q', \Sigma, \delta', s, \{f\})$, where $Q' = Q \cup \{r\}$ ($r \notin Q$), δ' is δ extended to include the new transition $r \in \delta(q, x)$. Let $q_1 \xrightarrow{a_1} q_2 \xrightarrow{a_2} \cdots \xrightarrow{a_{k-1}} q_k \xrightarrow{a_k} q_{k+1}$ be a labeled path in the transition diagram of A. *Deletion* of $a_1 a_2 \cdots a_{k-1}$ makes a new automaton $A' = (Q, \Sigma, \delta', s, \{f\})$, where δ' is δ extended to include the new transition $q_{k+1} \in \delta(q_1, a_k)$. The rank of the deletion is $k - 1$. Inserting or substituting k symbols has the rank equal to k. These error-correcting rules can be seen from a different point of view. Let w be a word we want to add to an automaton. According to the algorithm, in order to do it, we first look up a word u such that the Levenshtein distance[1] between u and

---

[1]Informally, the Levenshtein distance between two words is the minimum number of single-character edits (i.e. insertions, deletions or substitutions) required to change one word into the other.

© Springer International Publishing AG 2017
W. Wieczorek, *Grammatical Inference*, Studies in Computational Intelligence 673,
DOI 10.1007/978-3-319-46801-3_4

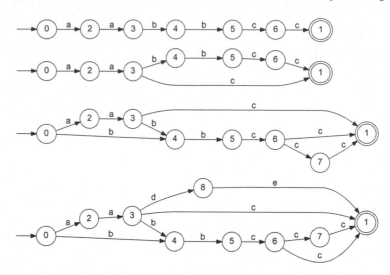

Fig. 4.1 From top to bottom, the automata after adding, respectively, aabbcc, aac, bbccc, and aade

w is minimal, and then do necessary operations. For example, if w = adf and u = abcde then it is sufficient to delete bc and substitute f for e. The overall rank of this transformation is equal to 3. But we have to be careful in keeping the initial and final states unchanged as well as in preserving the ingoing arcs property of an automaton.

An Example

Suppose that S_+ = {aabbcc, aac, bbccc, aade}. The automata that are obtained have been showed in Fig. 4.1. After putting aabbcc, the automaton has simple, linear structure. In order to process the word aac, only the deletion of bbc is required. Adding the next word, bbccc, needs the deletion of aa and insertion of c after 6th state. At this moment, the automaton stores the set {aabbcc, aac, bbccc, aabbccc, bbcc}. The 'closest' word to the aade word is aac. Thus, in order to add aade, a single insertion and a single substitution have to be performed.

The Formulation of Automata Induction

The induction scheme is as follows:

1. If $a_0a_1 \ldots a_i \ldots a_{n-1} \in S_+$ is the first word sample, then the initial automaton $A_0 = (Q, \Sigma, \delta, q_0, \{q_1\})$ is $Q = \{q_0, q_1, \ldots, q_n\}$, $\Sigma = \{a_0, a_1, \ldots, a_{n-1}\}$, $\delta(q_0, a_0) = \{q_2\}$, $\delta(q_i, a_{i-1}) = \{q_{i+1}\}$ for $i = 2, \ldots, n - 1$, $\delta(q_n, a_{n-1}) = \{q_1\}$.
2. For all training words $\beta = b_0 \ldots b_i \ldots b_l \in S_+$ do:

   a. With the current automaton, obtain the optimal (i.e., with the lowest rank) error-correcting path P of β, $P = p_0, e_0, p_1, e_1, p_2, \ldots, p_{k-1}, e_{k-1}, p_k$, where $p_i \in Q$ ($i = 0, 1, \ldots, k$), $p_0 = q_0$, $p_k = q_1$, and $e_j \in \Sigma$ or

$e_j \in \{d, i, s\} \times \Sigma$ ($j = 0, 1, \ldots k - 1$). The error rules dx, ix, and sx denote deletion of x, insertion of x, and substitution of x, respectively.

  b. For every i and j, $i < j$, such that e_i and e_j are two non-error rules of P between which there are (in P) only error rules, add to the current automaton the corresponding symbols and new states (skipping all dxs). The last added transition is $r \xrightarrow{e_j} p_{j+1}$, where r is a recently added state. If all error rules are deletions then only the transition $p_{i+1} \xrightarrow{e_j} p_{j+1}$ is added.

3. Return the current automaton.

4.1.2   Our Implementation

The minimal error-correcting rank is computed efficiently via the `memoize` decorator, which has been defined on page 122. Along with the rank, the C function determines the optimal error-correcting path as well.

```
from FAdo.fa import *

infinity = (float('inf'), ())

@memoize
def C(q, x, A):
  """Determine optimal error-correcting path
  Input: a state, a word, and an acyclic NFA
  Output: (rank, (q_0, e_0, p_1, e_1, ..., q_1))"""
  if x != "" and q not in A.Final:
    return min( \
      (1 + C(q, x[1:], A)[0], (q, 'i' + x[0]) + C(q, x[1:], A)[1]), \
      min( (C(r, x[1:], A)[0], (q, x[0]) + C(r, x[1:], A)[1]) \
      for r in A.delta[q][x[0]] ) if x[0] in A.delta[q] else infinity, \
      min( (1 + C(r, x[1:], A)[0], (q, 's' + x[0]) + C(r, x[1:], A)[1]) \
      for b in A.delta[q] for r in A.delta[q][b] ), \
      min( (1 + C(r, x, A)[0], (q, 'd' + b) + C(r, x, A)[1]) \
      for b in A.delta[q] for r in A.delta[q][b] ) )
  if x == "" and q not in A.Final:
    return infinity
  if x != "" and q in A.Final:
    return infinity
  if x == "" and q in A.Final:
    return (0, (q,))

def addPath(word, start, stop, A):
  """Inserts a word into an automaton between states start and stop
  Input: a string, two states, an acyclic NFA
  Outpu: an updated NFA"""
  i = start
  for c in word[:-1]:
    j = A.addState()
    A.addTransition(i, c, j)
    i = j
  A.addTransition(i, word[-1], stop)
```

```
def initial(a):
    """Builds an initial acyclic NFA
    Input: a word
    Output: an NFA"""
    A = NFA()
    init = A.addState()
    final = A.addState()
    A.addInitial(init)
    A.addFinal(final)
    addPath(a, init, final, A)
    return (A, init)

def synthesize(ListOfWords):
    """Synthesizes an acyclic NFA by means of the ECGI method
    Input: the list of example words
    Output: an NFA"""
    A, init = initial(ListOfWords[0])
    for idx in xrange(1, len(ListOfWords)):
        print A
        w = ListOfWords[idx]
        d = list(C(init, w, A)[1])
        print w, d
        n = len(d)
        i = 0
        while i < n-1:
            while i < n-1 and len(d[i+1]) == 1:
                i += 2
            j = i
            while j < n-1 and len(d[j+1]) == 2:
                j += 2
            if i < j:
                alpha = ""
                for k in xrange(i+1, j, 2):
                    if d[k][0] != "d":
                        alpha += d[k][1]
                if j < n-1:
                    addPath(alpha + d[j+1], d[i], d[j+2], A)
                else:
                    addPath(alpha, d[i], d[j], A)
            i = j+2
    return A
```

4.2 Alignment-Based Learning

Suppose that we are given three words: abcde, abccdf, and gbgdg. Alignment-Based Learning (ABL) is based on searching identical and distinct parts of words, which will be called *hypotheses*. For example, by comparing abcde with abccdf, we might conjecture the hypothesis

$$ab\{c, cc\}d\{e, f\},$$

but comparing abcde with gbgdg gives us the following hint:

$$\{a, g\}b\{c, g\}d\{e, g\}.$$

This phase of the present method is called alignment learning. During the alignment learning, when a hypothesis with the same structure of an existing hypothesis is found, then the two are merged (to be more precise, an existing hypothesis is updated). Since some obtained hypotheses may be overlapping or redundant, the second phase—selection learning—is performed. The aim of the selection learning phase is to select the correct hypotheses (or at least the better hypotheses). For our example words it is enough to yield a single hypothesis, which covers all the three examples:

$$\{a, g\}b\{c, cc, g\}d\{e, f, g\}.$$

This represents the following CFG:

$$S \rightarrow A\,b\,C\,d\,E \quad A \rightarrow a \mid g$$
$$C \rightarrow c \mid cc \qquad\quad E \rightarrow e \mid f \mid g$$

4.2.1 Alignment Learning

The alignment phase, i.e., linking identical fragments in the words, can be done in several different ways. The most common ones use the edit distance algorithm or the longest common subsequence. We will implement the latter approach. First, let us define a hypothesis as a sequence of words (identical parts) and sets of words (distinct parts, so-called constituents).

```
class Structure(object):
    """The class for storing hypotheses of Alignment-Based
    Learning"""

    def __init__(self, seq):
        """Structure is a list of sets (of strings)
        and strings"""
        self.__sequence = seq

    def __eq__(self, ob):
        if len(self.__sequence) != len(ob.__sequence):
            return False
        for (a, b) in zip(self.__sequence, ob.__sequence):
            if type(a) != type(b):
                return False
            if isinstance(a, str) and (a != b):
                return False
        return True

    def __nonzero__(self):
        return self.__sequence != []
```

```
def update(self, ob):
    """Adds new constituents to this structure
    Input: another compatible structure
    Output: an updated structure"""
    for (a, b) in zip(self.__sequence, ob.__sequence):
        if isinstance(a, set):
            a.update(b)

def size(self):
    """Returns the number of words represented by this object"""
    result = 1
    for el in self.__sequence:
        if isinstance(el, set):
            result *= len(el)
    return result

def words(self):
    """Returns the set of words represented by this structure"""
    return reduce(catenate, map( \
        lambda x: {x} if isinstance(x, str) else x,
        self.__sequence))
```

As can be seen from the listing, two hypotheses are compatible (the == operator) if their identical parts occupy the same places in sequences. Any hypothesis (an object of the Structure class) can be updated by extending its constituents using another compatible hypothesis. The size of a hypothesis is a number of words that the hypothesis represents.

Given a sequence $s_1, s_2, \ldots, s_n$ a *subsequence* is any sequence $s_{i_1}, s_{i_2}, \ldots, s_{i_m}$ with i_j strictly increasing. Let $z = z_1, z_2, \ldots, z_k$ be an lcs (the longest common subsequence) of $x = x_1, x_2, \ldots, x_m$ and $y = y_1, y_2, \ldots, y_n$. Then

- If $x_m = y_n$, then we must have $z_k = x_m = y_n$ and $z_1, z_2, \ldots, z_{k-1}$ is an lcs of $x_1, x_2, \ldots, x_{m-1}$ and $y_1, y_2, \ldots, y_{n-1}$.
- If $x_m \neq y_n$, then z uses at most one of them. Specifically:

  - If $z_k \neq x_m$, then z is an lcs of $x_1, x_2, \ldots, x_{m-1}$ and y.
  - If $z_k \neq y_n$, then z is an lcs of x and $y_1, y_2, \ldots, y_{n-1}$.

Treating words $x = x_1 x_2 \ldots x_m$ and $y = y_1 y_2 \ldots y_n$ as sequences $x = x_1, x_2, \ldots, x_m$ and $y = y_1, y_2, \ldots, y_n$, an lcs of x and y can be determined be the following efficient function:

```
def lcs(x, y):
    """Finds an lcs of two strings"""
    n = len(x)
    m = len(y)
    table = dict()
    for i in xrange(n+1):
        for j in xrange(m+1):
            if i == 0 or j == 0:
                table[i, j] = 0
            elif x[i-1] == y[j-1]:
                table[i, j] = table[i-1, j-1] + 1
            else:
                table[i, j] = max(table[i-1, j], table[i, j-1])
```

```
def recon(i, j):
  if i == 0 or j == 0:
    return ""
  elif x[i-1] == y[j-1]:
    return recon(i-1, j-1) + x[i-1]
    elif table[i-1, j] > table[i, j-1]:
    return recon(i-1, j)
  else:
    return recon(i, j-1)

return recon(n, m)
```

Based on the `lcs` function we can construct a function for finding a hypothesis from two words:

```
def align(x, y):
  """Finds the identical and distinct parts between two words
  Input: two strings
  Output: an instance of the Structure class"""
  seq = []
  same = lcs(x, y)
  i2, n = 0, len(same)
  ix, iy = 0, 0
  while i2 < n:
    wx, wy = "", ""
    i1 = i2
    while x[ix] != same[i1]:
      wx += x[ix]
      ix += 1
    while y[iy] != same[i1]:
      wy += y[iy]
      iy += 1
    while i2 < n and x[ix] == y[iy] == same[i2]:
      i2 += 1
      ix += 1
      iy += 1
    seq.append({wx, wy})
    seq.append(same[i1:i2])
  if same: seq.append({x[ix:], y[iy:]})
  return Structure(seq)
```

Finally, the whole alignment learning phase can be stated as follows.

1. Initiate an empty list of hypotheses.
2. For each pair (w_1, w_2) of different words:

   a. Align w_1 to w_2 and find the identical and distinct parts (a hypothesis) between them.
   b. Store a new or update an existing hypothesis.

3. Return the list of hypotheses.

4.2.2 Selection Learning

Not only can the alignment learning phase have different instantiations. Selection learning can also be implemented in different ways. There are two main approaches. In the first approach, we assume that the first hypothesis learned is the correct one. This means that when a new hypothesis overlaps with an older hypothesis, it can be ignored. In the second approach, we compute the probability of a hypothesis by counting the number of times the particular words of the hypothesis have occurred in other hypotheses. Herein, we propose another approach, in which all hypotheses are stored, but in the end we try to keep only the minimum number of hypotheses that cover all examples. Because set covering is an intractable problem, the greedy method has been used for selecting the subset of all obtained hypotheses. In our method, hypotheses that represent greater number of words are selected first.

4.2.3 Our Implementation

In the following closing function, the for-loop is responsible for the alignment learning and the while-loop is responsible for the selection learning.

```
def synthesize(S):
    """Finds a set of structures that covers S
    Input: a list of words
    Output: a list of structures"""
    n = len(S)
    hypotheses = []
    for i in xrange(n-1):
        for j in xrange(i+1, n):
            s = align(S[i], S[j])
            if s:
                k, m = 0, len(hypotheses)
                while k < m:
                    if s == hypotheses[k]:
                        hypotheses[k].update(s)
                        break
                    k += 1
                if k == m:
                    hypotheses.append(s)
    hypotheses.sort(key = lambda x: -x.size())
    Sp = set(S)
    X = hypotheses[0].words()
    idx = 1
    while not (Sp <= X):
        X.update(hypotheses[idx].words())
        idx += 1
    return hypotheses[:idx]
```

The function returns a list of hypotheses, which can be easily transformed into a non-circular[2] context-free grammar. This last step, called grammar extraction, is omitted, since language membership queries can be easily resolved through the `words` method.

4.3 Bibliographical Background

The error-correcting GI algorithm was introduced by Rulot and Vidal (1987). They presented the method in general terms, using a regular grammar. We have described it in details using the finite state automaton terminology. van Zaanen (2000) is the author of Alignment-Based Learning. As we could see from Sect. 4.2, this idea can be put into work using different building blocks. Another practically usable system is available at http://ilk.uvt.nl/menno/research/software/abl.

There are also two other representatives of substring-based GI concepts: ADIOS and Data-Oriented Parsing (DOP) approaches. ADIOS (Solan et al. 2005) starts by loading the corpus (examples) onto a directed graph whose vertexes are all lexicon entries, augmented by two special symbols, begin and end. Each corpus sentence defines a separate path over the graph, starting at begin and ending at end, and is indexed by the order of its appearance in the corpus. Loading is followed by an iterative search for significant patterns, which are added to the lexicon as new units. The algorithm generates candidate patterns by traversing in each iteration a different search path, seeking subpaths that are shared by a significant number of partially aligned paths. The significant patterns are selected according to a context-sensitive probabilistic criterion defined in terms of local flow quantities in the graph. At the end of each iteration, the most significant pattern is added to the lexicon as a new unit, the subpaths it subsumes are merged into a new vertex, and the graph is rewired accordingly. The search for patterns and equivalence classes and their incorporation into the graph is repeated until no new significant patterns are found.

The key idea of DOP (Bod 2006) is like this: given a sample, use all subtrees, regardless of size, of all binary parsing trees for sample words, to parse new words. In this model, we compute the probabilities of parse trees from the relative frequencies of the subtrees. Although it is now known that such a relative frequency estimator is statistically inconsistent, the model yields excellent empirical results and has been used in state-of-the-art systems.

---

[2]Let $G = (\{A_0, A_1, \ldots, A_m\}, \Sigma, P, A_0)$ be a CFG. A grammar G is said to be *non-circular* if the right-hand side of every A_i's rule does not contain a symbol A_j, where $j \leq i$. The non-circularity of a grammar guarantees that the language accepted by the grammar is finite.

References

Bod R (2006) An all-subtrees approach to unsupervised parsing. In: Proceedings of the 21st international conference on computational linguistics and 44th annual meeting of the ACL, association for computational linguistics, pp 865–872

Rulot H, Vidal E (1987) Modelling (sub)string-length based constraints through a grammatical inference method. In: Kittler J, Devijver P (eds) Proceedings of the NATO advanced study institute on pattern recognition theory and applications. Springer, pp 451–459

Solan Z, Horn D, Ruppin E, Edelman S (2005) Unsupervised learning of natural languages. Proc Nat Acad Sci USA 102(33):11,629–11,634

van Zaanen M (2000) ABL: alignment-based learning. In:Proceedings of the 18th international conference on computational linguistics (COLING), association for computational linguistics, association for computational linguistics, pp 961–967

Chapter 5
Identification Using Mathematical Modeling

5.1 From DFA Identification to Graph Coloring

This method has something in common with ideas described in Chap. 3. It starts
the searching of a minimum size automaton with generating a prefix tree acceptor,
then finds an appropriate partition of its states. However, the partition is obtained via
mathematical modeling.

5.1.1 Encoding

Definition 5.1 An *augmented* PTA (APTA) with respect to the positive part, S_+,
and the negative part, S_-, of a sample is a 6-tuple $(Q, \Sigma, \delta, q_0, F^+, F^-)$ where:

- $\text{PTA}(S_+ \cup S_-) = (Q, \Sigma, \delta, q_0, F^+ \cup F^-)$.
- F^+ and F^- are subsets of Q respectively identifying accepting states of S_+
  and S_-.
- F^+ is the set of final states of $\text{APTA}(S_+, S_-)$.

We give as an illustration (Fig. 5.1), the APTA of the sample $S_+ = \{\text{ab}, \text{aab}\}$,
$S_- = \{\text{a}, \text{ba}, \text{bb}\}$. States of F^+ are double lined and states of F^- are marked by
a dashed line. All remaining (neutral) states are enclosed by a single solid line.

By construction, the following properties hold: (1) $L(\text{APTA}(S_+, S_-)) = S_+$;
(2) $L(\text{APTA}(S_+, S_-)/\pi) = L(\text{PTA}(S_+)/\pi)$, where π is a partition of states of the
$\text{PTA}(S_+)$. Because we search for a deterministic automaton, only those partitions
will be considered that do not cause any non-determinism. The presence of F^- is
crucial, as it prevents the acceptance of negative words (no state $f^- \in F^-$ can be
merged with a final state $f^+ \in F^+$).

Now consider the graph whose vertexes are the states of an APTA and where there
is an edge between two vertexes (states) q and r if they cannot be merged. Then the

W. Wieczorek, *Grammatical Inference*, Studies in Computational Intelligence 673,
DOI 10.1007/978-3-319-46801-3_5

Fig. 5.1 An APTA
accepting ab, aab and
'rejecting' a, ba, bb

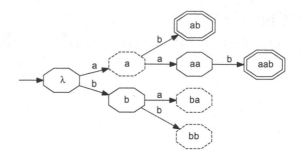

problem is to find a coloring[1] of the graph with a minimum number of colors. In fact, the minimum number of colors may be greater by one than the number of states in the smallest automaton. The reason given for this increase is the need to make an additional ('dead') state for some elements of F^- and possibly for some neutral states. For example, if $S_+ = \{\lambda, a, aa\}$, $S_- = \{b, bb\}$, then the smallest DFA has one state with a single transition (from the state to itself) labeled by a, but either of two states, for b and bb in an APTA, must not be merged with the accepting state. Thus, the present method returns in this case a two-state DFA.

There are two types of constraints which must be respected during graph building (the definition of the set of edges) and the coloring of its vertexes (the definition of a partition):

- Consistency constraints (on pairs of states): $q \in F^+$ cannot be merged with $r \in F^-$.
- Determinization constraints (on couple of pairs of states): if $\delta(q_1, a) = r_1$ and $\delta(q_2, a) = r_2$, then merging q_1 with q_2 implies that r_1 and r_2 must also be merged in order to produce a deterministic automaton.

5.1.2   Our Implementation

The above-mentioned approach can be easily implemented via CSP using modeling language described in Appendix C. Let $S = S_+ \cup S_-$ ($S_+ \cap S_- = \emptyset$, and let $APTA(S_+, S_-) = (Q, A, \delta, q_0, F, R)$. The integer variables will be x_q, where $q \in Q$. The value of x_q is m ($1 \leq m \leq |Q|$) if a state q belongs to the mth block in a resulting partition (coloring). Then, the required model can be written in OML as follows:

```
strModel = """Model[
  Parameters[Sets, A, Q],
  Parameters[Integers[0, 1], delta[Q, A, Q], F[Q], R[Q]],
```

---

[1]Graph coloring is an assignment of labels traditionally called 'colors' to elements of a graph subject to certain constraints. In its simplest form, which is employed here, it is a way of coloring the vertexes of a graph such that no two adjacent vertexes share the same color.

```
    Decisions[Integers[1, %d], x[Q]],
    Constraints[
      FilteredForeach[{p, Q}, {q, Q},
        F[p] == 1 & R[q] == 1, Unequal[x[p], x[q]]
      ],
      FilteredForeach[{a, A}, {i, Q}, {j, Q}, {k, Q}, {l, Q},
        delta[i, a, j] == 1 & delta[k, a, l] == 1,
          (x[i] == x[k]) -: (x[j] == x[l])
      ]
    ],
    Goals[Minimize[Max[Foreach[{q, Q}, x[q]]]]]
]"""
```

The rest of a program is responsible for feeding the model with concrete data, receiving the values of integer variables, and decoding the smallest DFA.

```
import clr
clr.AddReference('Microsoft.Solver.Foundation')
clr.AddReference('System.Data')
clr.AddReference('System.Data.DataSetExtensions')
from Microsoft.SolverFoundation.Services import *
from System.Data import DataTable, DataRow
from System.Data.DataTableExtensions import *
from System.IO import *
from FAdo.fa import *


def inducedDFA(P, A):
    """Join groups of states for a new automaton
    Input: the partition P (the list of frozensets) and an DFA A
    Output: a new DFA, A/P"""
    B = DFA()
    d = dict()
    K = range(len(A.States))
    for p in P:
        d[p] = j = B.addState(p)
        if A.Initial in p:
            B.setInitial(j)
        if p & A.Final:
            B.addFinal(j)
        for state in p:
            K[state] = j
    for q in A.delta:
        for a in A.delta[q]:
            r = A.delta[q][a]
            B.addTransition(K[q], a, K[r])
    return B


def buildAPTA(Sp, Sm):
    """Builds an augmented PTA
    Input: a sample
    Output: a DFA"""
    aut = buildPTA(Sp | Sm)
    i = -1
    for w in aut.States:
        i += 1
        if w in Sm:
            aut.delFinal(i)
    return aut


def xsToDFA(v, aut):
```

```
"""Builds a DFA from variables
Input: iterator over values (x[0] = value, x[1] = index)
Output: a DFA"""
t = dict()
for x in v:
  if int(x[0]) in t:
    t[int(x[0])].add(int(x[1]))
  else:
    t[int(x[0])] = {int(x[1])}
partition = map(frozenset, t.itervalues())
return inducedDFA(partition, aut)


def synthesize(S_plus, S_minus):
  """Finds a minimal DFA consistent with the input
  Input: a sample, S = (S_plus, S_minus)
  Output: a DFA"""
  apta = buildAPTA(S_plus, S_minus)
  context = SolverContext.GetContext()
  context.LoadModel(FileFormat.OML, StringReader(strModel % len(apta)))
  parameters = context.CurrentModel.Parameters
  for i in parameters:
    if i.Name == "delta":
      table = DataTable()
      table.Columns.Add("Q1", int)
      table.Columns.Add("A", str)
      table.Columns.Add("Q2", int)
      table.Columns.Add("Value", int)
      for q1 in xrange(len(apta)):
        for a in apta.Sigma:
          for q2 in xrange(len(apta)):
            if q1 in apta.delta and a in apta.delta[q1]
              and q2 == apta.delta[q1][a]:
              table.Rows.Add(q1, a, q2, 1)
            else:
              table.Rows.Add(q1, a, q2, 0)
      i.SetBinding[DataRow](AsEnumerable(table), "Value", "Q1", "A", "Q2")
    if i.Name == "F":
      table = DataTable()
      table.Columns.Add("Q", int)
      table.Columns.Add("Value", int)
      for q in xrange(len(apta)):
        table.Rows.Add(q, 1 if q in apta.Final else 0)
      i.SetBinding[DataRow](AsEnumerable(table), "Value", "Q")
    if i.Name == "R":
      table = DataTable()
      table.Columns.Add("Q", int)
      table.Columns.Add("Value", int)
      q = -1
      for w in apta.States:
        q += 1
        table.Rows.Add(q, 1 if w in S_minus else 0)
      i.SetBinding[DataRow](AsEnumerable(table), "Value", "Q")
  solution = context.Solve()
  for d in solution.Decisions:
    if d.Name == "x":
      aut = xsToDFA(d.GetValues(), apta)
  context.ClearModel()
  return aut
```

5.2 From NFA Identification to a Satisfiability Problem

A zero-one nonlinear programming problem (0–1 NP) deals with the question of whether there exists an assignment to binary variables $x = (x_1, x_2, \ldots, x_n)$ that satisfies the constraints $f_i(x) = 0, i \in I$, and (optionally) simultaneously minimizes (or maximizes) some expression involving x, where $f_i(x)$ $(i \in I)$ are given nonlinear functions. Binary variables $(x_i \in \{0, 1\})$ are usually used for such purposes as modeling yes/no decisions, enforcing disjunctions, enforcing logical conditions, etc. Below, the problem of NFA induction will be formulated, then it will be re-formulated as an INLP, then the last subsection will present an implementation which uses a SAT solver. Let Σ be an alphabet, let S_+ (examples) and S_- (counter-examples) be two finite sets of words over Σ, and let k be an integer. The goal of NFA induction is to determine a k-state NFA $A = (Q, \Sigma, \delta, s, F)$ such that $L(A)$ contains S_+ and is disjoint with S_-. In view of the task of finding a minimum size automaton, the following two facts are noticeable: (a) if there is no ℓ-state NFA consistent with S_+ and S_-, then for $1 \leq k < \ell$ there is no such k-state NFA either; (b) an integer x between 1 and B can be determined by $\lceil \log B \rceil$ questions of the form "Is $x > k$?" Thus, in order to obtain a minimum size NFA, it is enough to run the algorithm only a few times, owing to the fact that in practical situations we are dealing with automata possessing a dozen or so states. This, of course, can be accomplished by binary search.

5.2.1 Encoding

Let $S = S_+ \cup S_-$ (as usual $S_+ \cap S_- = \emptyset$), and let $P(S)$ be the set of all prefixes excluding the empty word of all words of S. The integer variables will be $x_{pq} \in \{0, 1\}$, $p \in P(S), q \in Q$; $y_{aqr} \in \{0, 1\}, a \in \Sigma, q, r \in Q$; and $z_q \in \{0, 1\}, q \in Q$. The value of x_{pq} is 1 if $q \in \delta(s, p)$ holds in an automaton A, $x_{pq} = 0$ otherwise. The value of y_{aqr} is 1 if $r \in \delta(q, a)$, $y_{aqr} = 0$ otherwise. Finally, we let $z_q = 1$ if $q \in F$ and zero if not. Let us now see how to describe the constraints of the relationship between an automaton A and a set S in terms of nonlinear equations and inequalities.

1. Naturally, according to the presence of the empty word we require that

$$
\begin{aligned}
z_s &= 1, & \lambda \in S_+, \\
z_s &= 0, & \lambda \in S_-.
\end{aligned}
$$

   One of the above equations is needed only if $\lambda \in S$. In the opposite case, the variable should not be settled in advance.
2. Every example has to be accepted by the automaton, but no counter-example should be. This can be written as

$$\sum_{q \in Q} x_{pq} z_q \geq 1, \qquad p \in S_+ - \{\lambda\},$$
$$\sum_{q \in Q} x_{pq} z_q = 0, \qquad p \in S_- - \{\lambda\}.$$

3. For $P(S) \ni p = a \in \Sigma$ we can have x_{pq} equal to 1 only in cases in which $q \in \delta(s, a)$; thus

$$x_{pq} - y_{psq} = 0, \qquad p \in \{a \in \Sigma : a \in P(S)\}, \ q \in Q.$$

4. Finally, we want to express the fact that whenever $x_{pq} = 1$ for $p = wa, w \in \Sigma^+$, $a \in \Sigma$, we have $q \in \delta(r, a)$ for at least one state r such that $x_{wr} = 1$. And vice versa, if a word w is spelled out by a path from s to a state r and there is a transition $r \xrightarrow{a} q$, then $x_{pq} = 1$ has to be fulfilled. We can guarantee this by requiring

$$-x_{pq} + \sum_{r \in Q} x_{wr} y_{arq} \geq 0, \qquad p \in \{wa \in P(S) : w \in \Sigma^+ \wedge a \in \Sigma\}.$$
$$x_{pq} - x_{wr} y_{arq} \geq 0, \qquad q, r \in Q.$$

It is worth emphasizing that in a computer implementation, every equation and every inequality from constraints (1)–(4) can be written efficiently by means of bit-wise operators, 'and,' 'or,' 'xor,' and the constants 0 and 1, without using multiplication, addition, subtraction, or comparison.

5.2.2　Our Implementation

In the light of the final remark from the previous subsection, the present approach can be easily implemented using SAT solving described in Appendix B.

```python
from satispy import Variable
from satispy.solver import Minisat
from operator import and_, or_

def encode(Sp, Sm, k):
  """Write NFA induction as SAT
  Input: examples, counter-examples, and an integer
  Output: boolean formula (Cnf), variables y and z"""
  idx = Idx(0)
  Sigma = alphabet(Sp | Sm)
  Q = range(k)
  P = prefixes(Sp | Sm)
  P.remove("")
  x = dict(((w, q), Variable(idx + 1)) for w in P for q in Q)
  y = dict(((a, p, q), Variable(idx + 1)) \
  for a in Sigma for p in Q for q in Q)
  z = dict((q, Variable(idx + 1)) for q in Q)
  st = [] # subject to (i.e. constraints)

  # The empty word inclusion
  if "" in Sp: st.append(z[0])
  if "" in Sm: st.append(-z[0])
```

```python
    # Acceptance of examples
    for w in Sp - {""}:
        st.append(reduce(or_, (x[w, q] & z[q] for q in Q)))

    # Rejection of counter-examples
    for w in Sm - {""}:
        for q in Q:
            st.append(-x[w, q] | -z[q])

    # Single-symbol prefixes inclusion
    for a in P & Sigma:
        for q in Q:
            st.append(-(x[a, q]    y[a, 0, q]))

    # Multi-symbol prefixes inclusion
    for w in P:
        if len(w) >= 2:
            v, a = w[:-1], w[-1]
            for q in Q:
                st.append(x[w, q] >> \
                reduce(or_, (x[v, r] & y[a, r, q] for r in Q)))
                for r in Q:
                    st.append((x[v, r] & y[a, r, q]) >> x[w, q])

    return (reduce(and_, st), y, z)

def decode(solution, y, z, k):
    """Constructs an NFA from the values of y, x and k > 0
    Input: satispy Solution and Variables
    Output: a k-state NFA"""
    A = NFA()
    for i in xrange(k):
        A.addState()
    A.addInitial(0)
    for q in xrange(k):
        if solution[z[q]]:
            A.addFinal(q)
    for ((a, p, q), var) in y.iteritems():
        if solution[var]:
            A.addTransition(p, a, q)
    return A

def synthesize(S_plus, S_minus, k):
    """Infers an NFA A consistent with the sample
    Input: the sets of examples and counter-examples, k > 0
    Output: a k-states NFA or None"""
    formula, y, z = encode(S_plus, S_minus, k)
    solver = Minisat()
    sol = solver.solve(formula)
    return decode(sol, y, z, k) if sol.success else None
```

5.3   From CFG Identification to a CSP

Let Σ be an alphabet, let S_+ (examples) and S_- (counterexamples) be two disjoint finite sets of words over Σ, and let $k > 0$ be an integer. The goal of CFG induction is to determine a grammar $G = (V, \Sigma, P, v_0)$ such that $L(G)$ contains S_+ and is disjoint from S_-. Moreover, the following criteria have to be fulfilled: G is in CNF, $|V| \le k$, and the sum of the lengths of the right hand sides of the rules is minimal (or maximal if instead of precision we want to achieve a good recall).

5.3.1   Encoding

Let $S = S_+ \cup S_-$ ($S_+ \cap S_- = \emptyset$, $\lambda \notin S$), and let F be the set of all proper factors of all words of S. An alphabet $\Sigma \subseteq F$ is determined by S, and variables V are determined by k: $V = \{v_0, v_1, \ldots, v_{k-1}\}$. The binary variables will be w_{if}, y_{ijl}, and z_{ia}, where $i, j, l \in V$, $a \in \Sigma$, and $f \in F$. The value of w_{if} is 1 if $i \Rightarrow^* f$ holds in a grammar G, and $w_{if} = 0$ otherwise. The value of y_{ijl} is 1 if $i \rightarrow jl \in P$, and $y_{ijl} = 0$ otherwise. Finally, we let $z_{ia} = 1$ if $i \rightarrow a \in P$ and zero if not. Let us now see how to describe the constraints of the relation between a grammar G and a set F in terms of nonlinear equations.

Naturally, every example has to be accepted by the grammar, but no counterexample should be. This can be written as

$$w_{v_0 s} = 1, \qquad s \in S_+,$$
$$w_{v_0 s} = 0, \qquad s \in S_-.$$

We want to express the fact that whenever $w_{if} = 1$ for $f = a$ or $f = bc$, $a \in \Sigma$, $b, c, f \in F$, we have at least one product $y_{ijl} \cdot w_{jb} \cdot w_{lc}$ or z_{ia} equal to 1. And vice versa, if, for instance, a factor $f = bc$ and there is a rule $i \rightarrow jl$, b can be derived from a variable j and c can be derived from a variable l, then $w_{if} = 1$ has to be fulfilled. We can guarantee this by requiring

$$w_{if} \;\leftrightarrow\; \sum_{j,l \in V,\, bc=f} y_{ijl} \cdot w_{jb} \cdot w_{lc} + (z_{if} \text{ if } f \in \Sigma)$$

for each $(i, f) \in V \times F$, where $\alpha \leftrightarrow \beta$ means if $\alpha = 0$ then $\beta = 0$ and if $\alpha = 1$ then $\beta \ge 1$. Obviously, we are to find the minimum value of the linear expression

$$2 \sum_{i,j,l \in V} y_{ijl} \;+\; \sum_{a \in \Sigma,\, i \in V} z_{ia}.$$

5.3.2 Our Implementation

Let us assume that a context-free grammar is represented as described on page 34. Once again we will take advantage of the OML.

```python
import clr
clr.AddReference('Microsoft.Solver.Foundation')
clr.AddReference('System.Data')
clr.AddReference('System.Data.DataSetExtensions')
from Microsoft.SolverFoundation.Services import *
from System.Data import DataTable, DataRow
from System.Data.DataTableExtensions import *
from System.IO import *
from pipetools import pipe

# OML Model string
strModel = """Model[
  // F = Factors built from S_+ (positives) and S_- (negatives)
  // {0, 1, ..., k-1} = K = grammar variables
  Parameters[Sets, F, K],
  // t[f] == 1 iff |f| == 1; u[a, b, c] == 1 iff ab == c
  Parameters[Integers[0, 1], t[F], u[F, F, F],
    positives[F], negatives[F], dummy[K]],

  // w[i, f] == 1 iff from i we can derive factor f
  // y[i, j, l] == 1 iff there is a rule i -> j l
  // z[i, a] == 1 iff there is a rule i -> a
  Decisions[Integers[0, 1], w[K, F], y[K, K, K], z[K, F]],

  // Minimize the number of grammar rules
  Goals[Minimize[FilteredSum[{i, K}, {a, F}, t[a] == 1, z[i, a]] +
    Sum[{i, K}, {j, K}, {l, K}, 2*y[i, j, l]]]],

  Constraints[
    // Every example can be derived from the start symbol 0
    FilteredForeach[{s, F}, positives[s] == 1, w[0, s] == 1],
    // None of the counterexamples can be derived from 0
    FilteredForeach[{s, F}, negatives[s] == 1, w[0, s] == 0],
    // There is no rule of the form i -> a a ...
    FilteredForeach[{s, F}, {i, K}, t[s] == 0, z[i, s] == 0],
    // Relation between y, z, and w
    Foreach[{i, K}, {f, F},
      w[i, f] == 1 -: (FilteredSum[{b, F}, {c, F}, {j, K}, {l, K},
        u[b, c, f] == 1, y[i, j, l]*w[j, b]*w[l, c]] + z[i, f]) >= 1],
    Foreach[{i, K}, {f, F},
      w[i, f] == 0 -: (FilteredSum[{b, F}, {c, F}, {j, K}, {l, K},
        u[b, c, f] == 1, y[i, j, l]*w[j, b]*w[l, c]] + z[i, f]) == 0]
  ]
]"""


def varsToCFG(y, z):
  """Builds a CFG from variables
  Input: iterator over iterators
  Output: a CFG"""
  g = set()
  for i in y:
    if int(i[0]):
      g.add(tuple(map(int, i[1:])))
```

```
  for i in z:
    if int(i[0]):
      g.add((int(i[1]), i[2]))
  return g

def synthesize(S_plus, S_minus, k):
  """Finds a CFG with k non-terminals consistent with the input
  Input: a sample, S = (S_plus, S_minus), an integer k
  Output: a CFG (productions as the set of tuples) or None"""
  factors = pipe | prefixes | suffixes
  F = factors(S_plus | S_minus)
  F.remove("")
  context = SolverContext.GetContext()
  context.LoadModel(FileFormat.OML, StringReader(strModel))
  parameters = context.CurrentModel.Parameters
  for i in parameters:
    if i.Name == "t":
      table = DataTable()
      table.Columns.Add("F", str)
      table.Columns.Add("Value", int)
      for f in F:
        table.Rows.Add(f, 1 if len(f) == 1 else 0)
      i.SetBinding[DataRow](AsEnumerable(table), "Value", "F")
    if i.Name == "u":
      table = DataTable()
      table.Columns.Add("A", str)
      table.Columns.Add("B", str)
      table.Columns.Add("C", str)
      table.Columns.Add("Value", int)
      for a in F:
        for b in F:
          for c in F:
            table.Rows.Add(a, b, c, 1 if a+b == c else 0)
      i.SetBinding[DataRow](AsEnumerable(table), "Value", \
      "A", "B", "C")
    if i.Name == "positives":
      table = DataTable()
      table.Columns.Add("F", str)
      table.Columns.Add("Value", int)
      for f in F:
        table.Rows.Add(f, 1 if f in S_plus else 0)
      i.SetBinding[DataRow](AsEnumerable(table), "Value", "F")
    if i.Name == "negatives":
      table = DataTable()
      table.Columns.Add("F", str)
      table.Columns.Add("Value", int)
      for f in F:
        table.Rows.Add(f, 1 if f in S_minus else 0)
      i.SetBinding[DataRow](AsEnumerable(table), "Value", "F")
    if i.Name == "dummy":
      table = DataTable()
      table.Columns.Add("K", int)
      table.Columns.Add("Value", int)
      for j in xrange(k):
        table.Rows.Add(j, 0)
      i.SetBinding[DataRow](AsEnumerable(table), "Value", "K")
  solution = context.Solve()
  if solution.Quality == SolverQuality.Optimal \
    or solution.Quality == SolverQuality.LocalOptimal:
```

```
for d in solution.Decisions:
    if d.Name == "y":
        y = d.GetValues()
    if d.Name == "z":
        z = d.GetValues()
    cfg = varsToCFG(y, z)
    context.ClearModel()
    return cfg
else:
    context.ClearModel()
    return None
```

5.4 Bibliographical Background

The transformation from DFA identification into graph coloring was proposed by Coste and Nicolas (1997). Their work was further developed by Heule and Verwer (2010), who presented an exact algorithm for identification of DFA which is based on SAT solvers.

Encoding from the transformation of NFA identification into SAT was taken from Wieczorek (2012). Jastrzab et al. (2016) gave an improved formulation of the problem of NFA induction by means of 0–1 nonlinear programming and proposed two parallel algorithms.

The most closely related work to CFG identification is by Imada and Nakamura (2009). Their work differs from ours in three respects. Firstly, they translated the learning problem for a CFG into an SAT, which is then solved by an SAT solver. We did the translation into 0–1 NP (zero-one nonlinear programming[2]), which is then solved by a CSP (constraint satisfaction programming) solver. Secondly, they minimize the number of rules, while we minimize the sum of the lengths of the rules. And thirdly, our encoding can be easily changed in such a way that every grammar we were to obtain would be unambiguous with respect to the examples, which is not the case in the comparable approach. The last point is essential in some applications, because they favor unambiguity. In this regard, a few exemplary applications can be found in a work by Wieczorek and Nowakowski (2016).

References

Coste F, Nicolas J (1997) Regular inference as a graph coloring problem. In: Workshop on grammar inference, automata induction, and language acquisition, ICML 1997

Heule M, Verwer S (2010) Exact DFA identification using SAT solvers. In: Grammatical inference: theoretical results and applications 10th international colloquium, ICGI 2010, Lecture notes in computer science, vol 6339. Springer, pp 66–79

---

[2]Our translation can be further re-formulated as an integer linear program, but the number of variables increases so much that this is not profitable.

Imada K, Nakamura K (2009) Learning context free grammars by using SAT solvers. In: Proceedings of the 2009 international conference on machine learning and applications, IEEE computer society, pp 267–272

Jastrzab T, Czech ZJ, Wieczorek W (2016) Parallel induction of nondeterministic finite automata. In: Parallel processing and applied mathematics, Lecture notes in computer science, vol 9573. Springer International Publishing, pp 248–257

Wieczorek W (2012) Induction of non-deterministic finite automata on supercomputers. JMLR Workshop Conf Proc 21:237–242

Wieczorek W, Nowakowski A (2016) Grammatical inference in the discovery of generating functions. In: Gruca A, Brachman A, Kozielski S, Czachórski T (eds) Man-machine interactions 4, advances in intelligent systems and computing, vol 391. Springer International Publishing, pp 627–637

Chapter 6
A Decomposition-Based Algorithm

6.1 Prime and Decomposable Languages

We start from the presentation of basic definitions, facts, and theorems for decomposability. The process of (multi)decomposing languages will be a key operation in algorithms described in this chapter.

6.1.1 Preliminaries

Graphs

A *graph G* is a finite non-empty set of objects called *vertexes* together with a (possibly empty) set of unordered pairs of distinct vertexes of G called *edges*. The vertex set of G is denoted by $V(G)$, while the edge set is denoted by $E(G)$. The edge $e = \{u, v\}$ is said to *join* the vertexes u and v. If $e = \{u, v\}$ is an edge of a graph G, then u and v are *adjacent vertexes*, while u and e are *incident*, as are v and e. Furthermore, if e_1 and e_2 are distinct edges of G incident with a common vertex, then e_1 and e_2 are *adjacent edges*. If $v \in V(G)$, then the set of vertexes adjacent to v in G is denoted by $N(v)$. The number $|N(v)|$, denoted by $d_G(v)$, is called the *degree* of a vertex v in a graph G.

In a graph G, a *clique* is a subset of the vertex set $C \subseteq V(G)$ such that every two vertexes in C are adjacent. If a clique does not exist exclusively within the vertex set of a larger clique then it is called a *maximal clique*. A *maximum clique* is a clique of the largest possible size in a given graph.

Language primality

A language is said to possess a *decomposition* if it can be written as a catenation of two languages neither one of which is the singleton language consisting of the empty word. Languages which are not such products are called *primes*. Thus, having given a decomposable (not prime) language L, we can determine both factors—such languages L_1, L_2 that $L = L_1L_2$, $L_1 \neq \{\lambda\}$ and $L_2 \neq \{\lambda\}$.

© Springer International Publishing AG 2017

W. Wieczorek, *Grammatical Inference*, Studies in Computational Intelligence 673,
DOI 10.1007/978-3-319-46801-3_6

In respect of finite languages the complexity of the operation of decomposing a language L into a non-trivial (we call $\{\lambda\}$ a trivial language) catenation $L_1 L_2$ is not well understood. Although for finite languages (every finite language is regular) the problem of primality is decidable, the decomposition of a finite language into primes may not be unique. Algorithmic issues of this problem are intriguing: it has been left open whether primality of a language L is NP-hard. Furthermore, there is no known algorithm which works on thousands of words of an arbitrary finite language in reasonable time. It has been suggested that in many cases, an exhaustive search is the only approach we can use.

6.1.2   Cliques and Decompositions

Let Σ be a finite non-empty alphabet and let $X \subset \Sigma \Sigma^+$ be a finite language. In this subsection we are going to show that there is a strong relation between the existence of a decomposition of X and a clique of size $|X|$ in a specially prepared graph.

Let $X = \{x_1, x_2, \ldots, x_n\}$ $(n \geq 1)$ be a given language with no words of size 0 (the empty word, λ) or 1. All possible word-splittings are denoted by $x_i = u_{ij} w_{ij}$, $i = 1, 2, \ldots, n$, $j = 1, 2, \ldots, |x_i| - 1$. The prefix u_{ij} consists of the j leading symbols of x_i, while the suffix w_{ij} consists of the $|x_i| - j$ trailing symbols of x_i. Consider the graph G with vertex set

$$V(G) = \bigcup_{i=1}^{n} \{(u_{ij}, w_{ij}) : j = 1, 2, \ldots, |x_i| - 1\}$$

and with edge set $E(G)$ given by:

$$\{(u_{ij}, w_{ij}), (u_{kl}, w_{kl})\} \in E(G) \ \Leftrightarrow \ i \neq k \ \wedge \ u_{ij} w_{kl} \in X \ \wedge \ u_{kl} w_{ij} \in X.$$

We will call G the *characteristic graph* of X.

Theorem 6.1 *Let G be the characteristic graph of X, $|X| = n$. A language X is decomposable if and only if there is a clique of size n in G.*

Proof (If part) Let $X = AB$ $(|X| = n)$, where A and B are non-trivial. Suppose that $A = \{a_1, a_2, \ldots, a_k\}$ and $B = \{b_1, b_2, \ldots, b_m\}$ $(1 \leq k, m \leq n)$. Naturally, every $a_i b_j$ must be a member of X and for every $x \in X$ there is at least one pair of i and j $(1 \leq i \leq k, 1 \leq j \leq m)$ such that $x = a_i b_j$. Let us construct a set C in the following way. For every $x \in X$ find such $a_i \in A$ and $b_j \in B$ that $x = a_i b_j$, then add the pair (a_i, b_j) to C. Observe that for every consecutive $x' = a_{i'} b_{j'}$ and any previously added $x = a_i b_j$, the pair $\{(a_i, b_i), (a_{i'}, b_{j'})\}$ is an edge in G. Thus, eventually $|C| = n$ and C is a clique in G.

(Only-if part) Now, we assume that $C = \{(a_1, b_1), (a_2, b_2), \ldots, (a_n, b_n)\}$ is a clique in G. From the definition of $E(G)$, we can conclude that for any $1 \leq i, k \leq n$,

$i \neq k$, the words $a_i b_i$ and $a_k b_k$ must be different; also we conclude that $a_i b_k$ and $a_k b_i$ are elements of X. Fix $A = \{a_1, a_2, \ldots a_n\}$ and $B = \{b_1, b_2, \ldots, b_n\}$. It is easily seen that $AB \supseteq \{a_1 b_1, a_2 b_2, \ldots a_n b_n\} = X$ (we have used only the facts that $a_i b_i$ are mutually different and X is of cardinality n). From the other side, $AB \subseteq X$ (since any $a_i b_k$ is an element of X). This clearly forces $X = AB$. $\square$

6.2 CFG Inference

The idea behind our GI algorithm inspired by the decomposition operation is as follows. We proceed along the general rule (which sometimes is not the best one) that states: the shorter description matching a sample the better hypothesis about the unknown language. So, according to this, we can apply the following greedy heuristic. At every stage of the inference process, try to do the decomposition on the largest possible subset of given data. In this way we achieve a small grammar that may be squeezed further by merging its nonterminals. The last phase has to be controlled by checking the rejection of counter-examples.

6.2.1 The GI Algorithm

Suppose we want to find a context-free grammar consistent with a finite sample $S \subset \Sigma^+$. In this section we discover a two-phase algorithm that does it efficiently.

Phase 1—constructing a grammar for the positive part of a sample

The language S_+ can be generated by a context-free grammar $G = (V, \Sigma, P, V_0)$ in flexible[1] Chomsky normal form. We can write:

$$V_0 \rightarrow a_1 \mid a_2 \mid \ldots \mid a_j \mid A B \mid C$$

where $a_i \in \Sigma$ are all single-symbol words of S_+, sets $A, B, C \subset \Sigma^+$, AB comes from the decomposition of some subset of $S_+ - \Sigma$, and C are remaining words of S_+. The decomposition is performed by building the characteristic graph of $S_+ - \Sigma$, searching the maximum clique in it, and determining sets A, B as described in the proof of Theorem 6.1. If the obtained clique has the (maximal) size $|S_+ - \Sigma|$, then $C = \emptyset$; in this case, a production $V_0 \rightarrow C$ is not present in the grammar.

Proceeding in the same way as described above, for sets A, B and C, and then recursively for the next sets, we obtain the grammar that generates exactly the language S_+. It finishes the first phase. The `initialG` function for obtaining this initial grammar is given in Sect. 6.2.2. Then, the grammar is transformed to Chomsky normal form by eliminating unit productions. In the second phase the size of the

---

[1] We call a CFG flexible if it allows unit productions ($A \rightarrow B$, $A, B \in V$). Let G be a grammar that contains unit productions. As it is well known, such a grammar G_1 can be constructed from G that $L(G_1) = L(G)$ and G_1 has no unit productions.

grammar is reduced by merging 'similar' nonterminals. The possibility of getting a grammar that generates an infinite language is the advantageous by-product of this reduction.

Phase 2—merging nonterminals

Now, we will describe the procedure which—in an iterative process—improves the current grammar by reducing the number of variables and consequently the number of rules. This reduction leads in turn to an increase in the number of accepted words and this is why, in some cases, we get a grammar that generates an infinite language.

Before the execution of $\texttt{InitialG}$, the set P and the map V (an associative array where a key is the set of words and a value is an integer index) are empty. To every set L in V is associated a number i ($V[L] = i$) such that a grammar variable V_i 'represents' the set L. Actually, $L = \{w \in \Sigma^+ \mid V_i \Rightarrow^* w\}$. $V[X] = 0$ and, thus, V_0 is the start symbol. Let $V[I] = i$ and $V[J] = j$. Then, those nonterminals V_i and V_j are merged first, that the set $I \cap J$ is largest. So as to avoid the acceptance of a word $w \in S_-$, after the mergence of nonterminals V_i and V_j, the improvement process is controlled with the help of the $\texttt{Parser}$ class defined on p. xx.

How can we be sure that the mergence of two variables V_i and V_j ($i < j$) only increases the number of accepted words? The skeleton of a proof would rely on a claim that if for any word $x \in \Sigma^+$ before the mergence $V_0 \Rightarrow^* u\,V_j\,w \Rightarrow^* x$ ($u, w \in (V \cup \Sigma)^*$) holds, then after the mergence we could write alike derivation, putting in the place of V_j a variable V_i.

6.2.2   Our Implementation

Let us assume that a context-free grammar is represented as described on pages 34 and 35. We have got the following form of the GI algorithm:

```python
import networkx as nx
from networkx.algorithms.approximation.clique import max_clique
from itertools import combinations
from operator import add

def buildCharacteristicGraph(X):
    """Builds the characteristic graph of a given language
    Input: set of strings, X
    Output: an networkx undirected graph"""
    V = dict((x, set()) for x in X)
    G = nx.Graph()
    for x in X:
        for i in xrange(1, len(x)):
            V[x].add((x[:i], x[i:]))
            G.add_node((x[:i], x[i:]))
    for (u, w) in combinations(G.nodes_iter(), 2):
        if u[0] + u[1] != w[0] + w[1] and u[0] + w[1] in X
        and w[0] + u[1] in X:
            G.add_edge(u, w)
    return G
```

```
def split(X):
  """Finds a split for a given language without the empty string
  nor words of length 1
  Input: a set of strings, X, epsilon not in X, and X is not
  the subset of Sigma
  Output: sets A, B, C for which AB + C = X and A, B are
  nontrivial"""
  G = buildCharacteristicGraph(X)
  K = max_clique(G)
  A = frozenset(u for (u, w) in K)
  B = frozenset(w for (u, w) in K)
  C = frozenset(X - catenate(A, B))
  return (A, B, C)

def initialG(X, V, P, T, idx):
  """Builds initial grammar (via V and P) for finite language X
  Input: a set of strings (without the empty string), X,
  references to an association table---variables V, productions P,
  an alphabet T, and consecutive index
  Output: an index for X (V, P, and idx are updated)"""
  i = idx[0]
  idx[0] += 1
  V[X] = i
  symbols = X & T
  if symbols:
    for s in symbols:
      P.add((i, s))
    X = X - symbols
  if X:
    A, B, C = split(X)
    a = V[A] if A in V else initialG(A, V, P, T, idx)
    b = V[B] if B in V else initialG(B, V, P, T, idx)
    P.add((i, a, b))
    if C:
      c = V[C] if C in V else initialG(C, V, P, T, idx)
      P.add((i, c))
  return i

def merge(i, j, P):
  """Joins two variables, i.e., V_j is absorbed by V_i
  Input: i, j variable indexes and grammar's productions P
  Output: a new production set, in which i and j have been
  merged into i"""
  Q = set()
  for p in P:
    Q.add(tuple(i if x == j else x for x in p))
  return Q

def eliminateUnitProductions(P):
  """Eliminates productions of the form A -> B (A, B in nonterminals)
  Input: the set of productions (tuples)
  Output: the set of productions without unit productions"""
  Q = set(filter(lambda p: len(p) == 3 or isinstance(p[1], str), P))
  U = P - Q
  if U:
    D = nx.DiGraph(list(U))
    paths = nx.shortest_path_length(D)
    V = set(reduce(add, U))
    for (a, b) in combinations(V, 2):
```

```
         if b in paths[a]:
           for p in P:
             if p[0] == b and (len(p) == 3 or isinstance(p[1], str)):
               Q.add((a,) + p[1:])
         if a in paths[b]:
           for p in P:
             if p[0] == a and (len(p) == 3 or isinstance(p[1], str)):
               Q.add((b,) + p[1:])
     return Q


def makeCandidateVariablesList(V):
    """Builds the sorted list of pairs of variables to merge
    Input: a dictionary of variables, V
    Output: a list of pairs of variables' indexes,
    first most promising"""
    card = dict()
    pairs = []
    for (I, i) in V.iteritems():
      for (J, j) in V.iteritems():
        if i < j:
          card[i, j] = len(I & J)
          pairs.append((i, j))
    pairs.sort(key = lambda x: -card[x])
    return pairs


def simplifiedGrammar(P, S):
    """Removes unnecessary productions
    Input: the set of productions, S_plus
    Output: the updated set of productions"""
    if len(P) > 1:
      cpy = list(P)
      for prod in cpy:
        P.remove(prod)
        parser = Parser(P)
        if not all(parser.accepts(w) for w in S):
          P.add(prod)
    return P


def synthesize(S_plus, S_minus):
    """Infers a CFG consistent with the sample
    Input: the sets of examples and counter-examples
    (without the empty string)
    Output: a CFG in CNF"""
    V = dict()
    P = set()
    T = alphabet(S_plus)
    idx = [0]
    initialG(frozenset(S_plus), V, P, T, idx)
    P = eliminateUnitProductions(P)
    joined = True
    while joined:
      pairs = makeCandidateVariablesList(V)
      joined = False
      for (i, j) in pairs:
        Q = merge(i, j, P)
        parser = Parser(Q)
        if not any(parser.accepts(w) for w in S_minus):
          P = Q
          for (key, val) in V.iteritems():
```

```
      if val == i: I = key
      if val == j: J = key
   del V[I]
   del V[J]
   V[I | J] = i
   joined = True
   break
return simplifiedGrammar(P, S_plus)
```

6.3 Bibliographical Background

Finite language decomposition was investigated for the first time by Mateescu et al. (1998). They discovered the concept of a decomposition set—the subset of the states of the minimal finite deterministic automaton for a regular language (notice that every finite language is regular). In their work, it was also left open whether testing the primality of a finite language is an intractable problem. Wieczorek (2010a) developed this theory and introduced the concept of significant states in DFA. It helped to construct a very clever algorithm for finding decompositions. In the paper we can find two additional algorithms; one of the algorithms is based on a graph which is very similar to the characteristic graph defined in the present chapter. Jastrzab et al. (2016) showed how the main algorithm in Wieczorek (2010a) may be parallelized. Wieczorek (2009) proposed also a few metaheuristic approaches for the decomposability of finite languages.

The problem of obtaining languages applying an operation on smaller languages is also crucial to the theory of formal languages (Ito 2004). Apart from catenation (i.e. the inverse operation to decomposition), the shuffle operation has been intensively studied in formal language theory (Berstel and Boasson 2002). Jedrzejowicz and Szepietowski (2001a) investigated the complexity of languages described by some expressions containing shuffle operator and intersection. The same authors considered the class of shuffle languages which emerges from the class of finite languages through regular operations (union, catenation, Kleene star) and some shuffle operations (Jedrzejowicz and Szepietowski 2001b).

Constructing an initial grammar in the algorithm described in this chapter (the initialG function) is taken from Wieczorek (2010b). A function for eliminating unit productions is an implementation of the algorithm given by Hopcroft et al. (2001).

References

Berstel J, Boasson L (2002) Shuffle factorization is unique. Theor Comput Sci 273(1–2):47–67
Hopcroft JE, Motwani R, Ullman JD (2001) Introduction to automata theory, languages, and computation, 2nd edn. Addison-Wesley
Ito M (2004) Algebraic theory of automata and languages. World Scientific

Jastrzab T, Czech Z, Wieczorek W (2016) A parallel algorithm for decomposition of finite languages. In: Parallel computing: on the road to exascale, advances in parallel computing, vol 27. IOS Press, pp 401–410

Jedrzejowicz J, Szepietowski A (2001a) On the expressive power of the shuffle operator matched with intersection by regular sets. Theor Inf Appl 35(4):379–388

Jedrzejowicz J, Szepietowski A (2001b) Shuffle languages are in P. Theor Comput Sci 250(1–2):31–53

Mateescu A, Salomaa A, Yu S (1998) On the decomposition of finite languages. Technical Report 222, Turku Centre for Computer Science

Wieczorek W (2009) Metaheuristics for the decomposition of finite languages. In: Recent advances in intelligent information systems. Academic Publishing House EXIT, pp 495–505

Wieczorek W (2010a) An algorithm for the decomposition of finite languages. Logic J IGPL 18(3):355–366

Wieczorek W (2010b) A local search algorithm for grammatical inference. In: Grammatical inference: theoretical results and applications, lecture notes in computer science, vol 6339. Springer, pp 217–229

Chapter 7
An Algorithm Based on a Directed Acyclic Word Graph

7.1 Definitions

A *directed graph*, or *digraph*, is a pair $G = (V, A)$, where V is a finite, non-empty set of *vertexes* and A has as elements ordered pairs of different vertexes called *arcs*; that is, $A \subseteq V \times V$. In a digraph $G = (V, A)$ the *in-degree* of a vertex v is the number of arcs of the form (u, v) that are in A. Similarly, the *out-degree* of v is the number of arcs of A that have the form (v, u). A *walk* $w = (v_1, v_2, \ldots, v_k)$ of G is a sequence of vertexes in V such that $(v_j, v_{j+1}) \in A$ for $j = 1, \ldots, k-1$. Furthermore, if $k > 1$ and $v_k = v_1$, then w is said to be *closed*. A *path* in G is a walk without repetitions. A *cycle* is a closed path.

We denote as usual by Σ^* the set of words over Σ and by Σ^+ the set $\Sigma^* - \{\lambda\}$. The set of all prefixes that can be obtained from the set of words X will be denoted by $P(X)$. Let X be a subset of Σ^*. For $w \in \Sigma^*$, we define the *left quotients* $w^{-1}X = \{u \in \Sigma^* \mid wu \in X\}$.

A DAWG $G = (s, t, V, A, \Sigma, \ell)$ is a digraph (V, A) with no cycles together with an alphabet Σ and with a label $\ell(u, v) \in 2^{\Sigma} - \{\emptyset\}$ for each $(u, v) \in A$, in which there is exactly one vertex with in-degree 0—the *source s*, and exactly one vertex with out-degree 0—the *terminal t*. It can be proved that in any DAWG, every vertex is reachable from the source. Furthermore, from every vertex the terminal is reachable. So there are no useless vertexes. We will say that a word $w \in \Sigma^+$ is *stored* by G if there is a labeled path from the source to the terminal such that this path spells out the word w. Let $L(G)$ be the set of all words that are spelled out by paths from s to t. We can formally define it by the *transition function* $\delta : V \times \Sigma^+ \to 2^V$ which is given inductively by

1. $\delta(v, a) = \{u \in V \mid (v, u) \in A \wedge a \in \ell(v, u)\}$, for $a \in \Sigma$,
2. $\delta(v, wa) = \bigcup_{u \in \delta(v, w)} \delta(u, a)$, for $w \in \Sigma^+$ and $a \in \Sigma$.

Therefore, we have $L(G) = \{w \in \Sigma^+ : t \in \delta(s, w)\}$. Using the terminology of automata, a DAWG could be referred to as a non-deterministic finite automaton that has exactly one accepting state and recognizes a finite language.

© Springer International Publishing AG 2017
W. Wieczorek, *Grammatical Inference*, Studies in Computational Intelligence 673,
DOI 10.1007/978-3-319-46801-3_7

Let $G = (s, t, V, A, \Sigma, \ell)$ be a directed acyclic word graph. To measure the *potency*, p, of $(u, v) \in A$ we make the following definition: $p(u, v)$ is the number of paths from v to t or 1 if $v = t$. Naturally, for all $v \in V - \{s, t\}$ we have $p(u, v) = \sum_{(v,w) \in A} p(v, w)$. Thus, by means of memoization, all potencies can be determined in time $O(|A|^2)$.

7.2   Constructing a DAWG From a Sample

Here we assume that a sample $S = (S_+, S_-)$ does not contain the empty word. The present algorithm is two-phased. In the first phase, based on the positive part of S, an initial DAWG, G_{S_+}, is constructed. Its aim is to discover the structure of an unknown finite language. In the second phase the DAWG is extended step by step by the addition of new labels. The possibility of getting a DAWG that stores an expected language is the purpose of this extension procedure. So as to avoid the acceptance of a counter-example, the second phase is controlled by means of S_-.

An initial DAWG is constructed on the basis of the following definitions. Its set of vertexes is

$$V = \{w^{-1}S_+ \mid w \in P(S_+)\}.$$

Its source is $s = S_+$, its terminal is $t = \{\lambda\}$, and its arcs and labels are defined for $v \in V$ and $a \in \Sigma$ by $a \in \ell(v, a^{-1}v)$ iff $a^{-1}v \neq \emptyset$ and $a \in \ell(v, t)$ iff $a \in v$. From these definitions we can conclude that $L(G_{S_+}) = S_+$. For example, the initial DAWG $G_{\{aba,baa,b\}}$ is shown in Fig. 7.1. Needless to say, in a computer implementation after building a digraph it is more convenient to represent its vertexes as integers rather than sets of words.

We are now in a position to describe the extend method, which, in an iterative process, improves the current word digraph by putting some additional labels onto the existing arcs. This auxiliary deposal leads in turn to an increase in the number of stored words and this is why we can get a DAWG that represents a more accurate or exact (sought) language. In the method, for each arc and for each symbol in Σ, the symbol is put onto the arc unless some counter-examples are accepted by the discussed DAWG. Please notice that the order of putting new labels alters the results, hence a greedy heuristic is used in order to obtain the most words consistent with a sample. The algorithm starts with the arcs of higher potency, then those of low potency are processed. The language membership problem (test if $t \in \delta(s, w)$)

Fig. 7.1 The initial DAWG built based on the set {aba, baa, b}

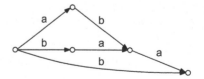

is efficiently resolved by the `accepts` method derived from the definition of δ in time $O(|w| \cdot |V|^2)$.

Let n be the sum of the lengths of all examples, m be the sum of the lengths of all counter-examples, and k be the size of an alphabet. Since both $|V|$ and $|A|$ in G_{S_+} are $O(n)$, and the computation cost of determining V and A is in $O(n^2)$, the running time of the present method is $O(kmn^3)$. It is also noteworthy that an achieved DAWG stores only words of the same length as those observed in the set of examples. Therefore, it is pointless to have a negative example of length i, when none of the positive examples is of length i.

7.3 Our Implementation

Below we can find the definition of the DAWG class. Its usage is very simple. First, build an initial DAWG using the positive part of a sample and a constructor, then extend the DAWG using the negative part of a sample and a proper method.

```
from itertools import count
import networkx as nx

class DAWG(object):
    """A class for storing DAWGs for the GI algorithm"""

    def __init__(self, Sp):
        """Constructs an initial DAWG from given words
        Input: the set of strings (not containing the empty word)
        Output: this object becomes the initial DAWG"""

        def lq(w, X):
            n = len(w)
            U = set()
            for x in X:
                if x[:n] == w:
                    U.add(x[n:])
            return frozenset(U)

        def isArc(a, u, v):
            if v == {''}:
                return a in u
            for x in u:
                if x != '' and x[0] == a and x[1:] not in v:
                    return False
            for x in v:
                if a+x not in u:
                    return False
            return True

        self.__Sigma = alphabet(Sp)
        P = prefixes(Sp)
        V = dict(zip(set(lq(w, Sp) for w in P), count(0)))
        self.__s = V[frozenset(Sp)]
        self.__t = V[frozenset({''})]
```

```python
    self.__ell = dict()
    for a in self.__Sigma:
      for u in V:
        for v in V:
          if isArc(a, u, v):
            ind = (V[u], V[v])
            if ind in self.__ell:
              self.__ell[ind].add(a)
            else:
              self.__ell[ind] = {a}

def accepts(self, word):
  """Checks whether given word is stored in this DAWG
  Input: a string
  Output: True or False"""

  def delta(q, w):
    if (q, w) in cache:
      return cache[q, w]
    else:
      if w == "" and q == self.__t:
        cache[q, w] = True
        return True
      elif w == "" and q != self.__t:
        cache[q, w] = False
        return False
      else:
        for (u, v) in self.__ell:
          if u == q and w[0] in self.__ell[u, v]:
            if delta(v, w[1:]):
              cache[q, w] = True
              return True
        cache[q, w] = False
        return False

  cache = dict()
  return delta(self.__s, word)

def __findPotencyList(self):
  """Finds a potency for every arc
  Input: this object
  Output: sorted arcs according to decreasing potencies"""
  D = nx.DiGraph(self.__ell.iterkeys())
  res = dict()
  for (u, v) in self.__ell.iterkeys():
    if v == self.__t:
      res[u, v] = 1
    else:
      res[u, v] = len(list( \
        nx.all_simple_paths(D, source=v, target=self.__t)))
  return list(reversed(sorted(res, key=res.__getitem__)))

def extend(self, Sm):
  """The second phase of the GI algorithm
  Input: the set of counter-examples
  Output: this DAWG is updated"""
  A = self.__findPotencyList()
```

```
for (u, v) in A:
  for a in self.__Sigma:
    if a not in self.__ell[u, v]:
      self.__ell[u, v].add(a)
      if any(self.accepts(w) for w in Sm):
        self.__ell[u, v].remove(a)
```

7.4 Bibliographical Background

With respect to the problem of DAWG construction one should pay attention to a number of material facts. There are also two types of DAWGs (as for the finite state automata). A directed graph is called deterministic when no transitions exist that have the same labels and leave the same state. This property results in a very efficient search function. Graphs that do not have this property are called nondeterministic. The latter are generally smaller than the former but they are a little slower to search. Throughout the chapter, we assumed a DAWG to be nondeterministic. It is well known that NFA or a regular expression minimization (which are of a nondeterministic character, too) is computationally hard: it is PSPACE-complete (Meyer and Stockmeyer 1972). Jiang and Ravikumar (1993) showed, moreover, that the minimization problem for NFAs or regular expressions remains PSPACE-complete, even when specifying the regular language by a DFA. Thus the problem of constructing a k-vertex DAWG that matches a set of input words is probably of exponential complexity. Some work has been done on the problem but for the set of examples only (i.e., without the presence of counter-examples and without the aim of generalization), see the algorithms devised by Amilhastre et al. (1999) and Sgarbas et al. (2001).

The present description of the algorithm based on a DAWG is in the most part taken from the work of Wieczorek and Unold (2014), in which the authors started applying it to a real bio-informatics task, i.e., the classification of amyloidogenic hexapeptides.

References

Amilhastre J, Janssen P, Vilarem M (1999) FA minimisation heuristics for a class of finite languages. In: Lecture notes in computer science, vol 2214. Springer, WIA, pp 1–12

Jiang T, Ravikumar B (1993) Minimal NFA problems are hard. SIAM J Comput 22:1117–1141

Meyer AR, Stockmeyer LJ (1972) The equivalence problem for regular expressions with squaring requires exponential space. In: Proceedings of the 13th annual symposium on switching and automata theory, pp 125–129

Sgarbas KN, Fakotakis N, Kokkinakis GK (2001) Incremental construction of compact acyclic NFAs. In: ACL. Morgan Kaufmann Publishers, pp 474–481

Wieczorek W, Unold O (2014) Induction of directed acyclic word graph in a bioinformatics task. JMLR Workshop Conf Proc 34:207–217

Chapter 8
Applications of GI Methods in Selected Fields

8.1 Discovery of Generating Functions

This section will be especially concerned with explaining how the induction of
a context-free grammar can support discovering ordinary generating functions (also
called OGFs). In combinatorics, the closed form of an OGF is often the basic way
of representing infinite sequences. Suppose that we are given a description of the
construction of some structures we wish to count. The idea is as follows. First, define
a one-to-one correspondence (a bijection) between the structures and the language
over a fixed alphabet. Next, determine some examples and counter-examples. Infer
an unambiguous context-free grammar consistent with the sample. Via classical
Schützenberger methodology, give a set of function equations. Finally, solve it and
establish a hypothesis for the OGF.

Our main contribution is to provide a procedure for inferring a small (by minimiz-
ing $|V|$) context-free grammar which accepts all examples, none of the counterex-
amples, and is unambiguous with respect to the examples. The grammar is likely
to be unambiguous in general, which is crucial in the Schützenberger methodology.
The most closely related work to this study is Sect. 5.3. We will prepare a program
similar to the one given there and use it to our goal.

This section is organized into three subsections. Sect. 8.1.1 introduces the notion of
ordinary generating functions. In Sect. 8.1.2 the relation between grammars and generating functions is explained. Finally, Sect. 8.1.3 discusses the experimental results.

8.1.1 Generating Functions

Let $a_0, a_1, \ldots$ be a sequence (especially of non-negative integers). Then the power
series

$$A(z) = \sum_{i=0}^{\infty} a_i z^i$$

© Springer International Publishing AG 2017

W. Wieczorek, *Grammatical Inference*, Studies in Computational Intelligence 673,
DOI 10.1007/978-3-319-46801-3_8

is called the *ordinary generating function* (OGF) associated with this sequence. For simple sequences, the closed forms of their OGFs can be obtained fairly easily. Take the sequence 2, 1, 2, 1, 2, ..., for instance:

$$A(z) = 2 + z + 2z^2 + z^3 + 2z^4 + \cdots = 2 + z + z^2(2 + z + 2z^2 + \cdots).$$

After a few transformations we see that $A(z) = (2 + z)/(1 - z^2)$. There are many tools from algebra and calculus to obtain information from generating functions or to manipulate them. Partial fractions are a good case in point, since they have been shown to be valuable in solving many recursions.

Why is the generating function one of the prime methods for the representation of infinite sequences? Because we can retrieve any a_i ($i \geq 0$) from the compact form of $A(z)$ by expanding $A(z)$ in a power series, using Maclaurin's formula:

$$A(z) = A(0) + \frac{z}{1!}A'(0) + \frac{z^2}{2!}A''(0) + \cdots + \frac{z^n}{n!}A^{(n)}(0) + \cdots.$$

It can be done automatically by means of such well-known computer algebra systems as Maxima or Mathematica.

8.1.2 The Schützenberger Methodology

We assume that $G = (V, \Sigma, P, S)$ is a context-free unambiguous grammar, and a_i is the number of words of length i in $L(G)$. Let Θ be a map that satisfies the following conditions:

1. for every $a \in \Sigma$, $\Theta(a) = z$,
2. $\Theta(\lambda) = 1$,
3. for every $N \in V$, $\Theta(N) = N(z)$.

If for every set of rules $N \to \alpha_1 \mid \alpha_2 \mid \ldots \mid \alpha_m$ we write $N(z) = \Theta(\alpha_1) + \Theta(\alpha_2) + \cdots + \Theta(\alpha_m)$, then

$$S(z) = \sum_{i=0}^{\infty} a_i z^i.$$

Thus, to take one example, let us try to find the number of ways in which one can put two indistinguishable balls into n distinguishable boxes. E.g., for three boxes there are six ways: [O] [O] [], [O] [] [O], [] [O] [O], [OO] [] [], [] [OO] [], and [] [] [OO]. It is easy to see that every valid throwing is accepted by the grammar

$$A \to B [OO] B \mid B [O] B [O] B$$
$$B \to \lambda \mid [] B$$

This clearly forces $A(z) = z^4 B^2(z) + z^6 B^3(z)$ and $B(z) = 1 + z^2 B(z)$, and consequently $A(z) = z^4/(1 - z^2)^2 + z^6/(1 - z^2)^3$. The Maclaurin series coefficients of $A(z)$ determine the number of words of a fixed length, so we can look up the number of unique throwings: $\langle 0, 0, 0, 0, 1, 0, 3, 0, 6, \ldots \rangle$ for consecutive lengths $n = 0, 1, 2, \ldots$. Because for k boxes the length of an associated word is $2k + 2$, we get rid of odd-positioned zeros and then shift $A(z)$ to the left by one place. It can be done through substituting z for z^2 and dividing $A(z)$ by z.[1] Finally, the generating function we are looking for is:

$$\frac{z}{(1 - z)^2} + \frac{z^2}{(1 - z)^3} = z + 3z^2 + 6z^3 + 10z^4 + 15z^5 + 21z^6 + \cdots .$$

8.1.3 Applications

In this section, four examples of problems from enumerative combinatorics will be solved by means of our GI method. In every problem, we are given an infinite class of finite sets S_i where i ranges over some index set I, and we wish to count the number $f(i)$ of elements of each S_i "simultaneously," i.e., give a generating function. In the first example we will define the necessary bijections on our own and show the whole process of reaching an OGF. The remaining examples will only be briefly reported.

Routes in a grid city

Let us try to find the number of ways in which one can go from point $(0, 0)$ to point (n, n) in the Cartesian plane using only the two types of moves: $u = (0, 1)$ and $r = (1, 0)$. So as to get an OGF that 'stores' these numbers for $n = 0, 1, 2, \ldots$, we first write a program for inferring a grammar in quadratic Greibach normal form with the minimal number of non-terminals, next give a procedure for using it in a way that minimizes its running time while maximizing a chance that an obtained grammar is unambiguous, then define a bijection between routes and words, and finally apply the three routines in cooperation.

Definition 8.1 A CFG $G = (V, \Sigma, P, S)$ is said to be in *quadratic Greibach normal form* if each of its rules is in one of the three possible forms:

1. $A \to a,\quad a \in \Sigma, A \in V,$
2. $A \to aB,\quad a \in \Sigma, A, B \in V,$
3. $A \to aBC,\quad a \in \Sigma, A, B, C \in V.$

---

[1] To shift $A(z)$ to the left by m places, that is, to form the OGF for the sequence $\langle a_m, a_{m+1}, a_{m+2}, \ldots \rangle$ with the first m elements discarded, we subtract the first m terms and then divide by z^m:

$$\frac{A(z) - a_0 - \cdots - a_{m-1} z^{m-1}}{z^m} = \sum_{n \geq m} a_n z^{n-m} = \sum_{n \geq 0} a_{n+m} z^n .$$

A below-presented GI program is based on the same principle of encoding like this one given in Sect. 5.3. So we will not repeat that explanation. There are, however, two important differences. Now, on output a grammar in quadratic Greibach normal form is expected. What is more, this grammar is unambiguous with respect to examples.

```python
import clr
clr.AddReference('Microsoft.Solver.Foundation')
clr.AddReference('System.Data')
clr.AddReference('System.Data.DataSetExtensions')
from Microsoft.SolverFoundation.Services import *
from System.Data import DataTable, DataRow
from System.Data.DataTableExtensions import *
from System.IO import *
from pipetools import pipe


# OML Model string
strModel = """Model[
  // F = Factors built from S_+ (positives) and S_- (negatives)
  // {0, 1, ..., k-1} = K = grammar variables
  Parameters[Sets, F, K],
  // t[f] == 1 iff |f| == 1; u[a, b, c] == 1 iff ab == c;
  // v[a, b, c, d] == 1 iff abc == d
  Parameters[Integers[0, 1], t[F], u[F, F, F], v[F, F, F, F],
    positives[F], negatives[F], dummy[K]],

  // w[i, f] == 1 iff from i we can derived factor f
  // x[i, a, j, k] == 1 iff there is a rule i -> a j k
  // y[i, a, j] == 1 iff there is a rule i -> a j
  // z[i, a] == 1 iff there is a rule i -> a
  Decisions[Integers[0, 1], w[K, F], x[K, F, K, K], y[K, F, K],
    z[K, F]],

  // Minimize the number of grammar rules
  Goals[Minimize[FilteredSum[{i, K}, {a, F}, t[a] == 1, z[i, a]] +
    FilteredSum[{i, K}, {a, F}, {j, K}, t[a] == 1, 2*y[i, a, j]] +
      FilteredSum[{i, K}, {a, F}, {j, K}, {k, K}, t[a] == 1,
        3*x[i, a, j, k]]]],

  Constraints[
    // Every example can be derived from the start symbol 0
    FilteredForeach[{s, F}, positives[s] == 1,  w[0, s] == 1],
    // None of the counterexamples can be derived from 1
    FilteredForeach[{s, F}, negatives[s] == 1,  w[0, s] == 0],
    // There is no rule of the form i -> a a ...
    FilteredForeach[{s, F}, {i, K}, t[s] == 0, z[i, s] == 0],
    // Relation between x, y, z, and w
    Foreach[{k, K}, {f, F},
      w[k, f] == FilteredSum[{a, F}, {b, F}, {c, F}, {i, K}, {j, K},
        v[a, b, c, f] == 1, x[k, a, i, j]*w[i, b]*w[j, c]] +
          FilteredSum[{a, F}, {b, F}, {i, K}, u[a, b, f] == 1,
            y[k, a, i]*w[i, b]] + z[k, f]]
  ]
]"""


def varsToCFG(*v):
    """Builds a CFG from variables
    Input: iterator over iterators
    Output: a CFG"""
    g = set()
```

```
  for it in v:
    for i in it:
      if int(i[0]):
        g.add((int(i[1]), i[2]) + tuple(map(int, i[3:]))))
  return g

def synthesize(S_plus, S_minus, k):
  """Finds a CFG with k non-terminals consistent with the input
  Input: a sample, S = (S_plus, S_minus), an integer k
  Output: a CFG (productions as the set of tuples) or None"""
  factors = pipe | prefixes | suffixes
  F = factors(S_plus | S_minus)
  F.remove("")
  context = SolverContext.GetContext()
  context.LoadModel(FileFormat.OML, StringReader(strModel))
  parameters = context.CurrentModel.Parameters
  for i in parameters:
    if i.Name == "t":
      table = DataTable()
      table.Columns.Add("F", str)
      table.Columns.Add("Value", int)
      for f in F:
        table.Rows.Add(f, 1 if len(f) == 1 else 0)
      i.SetBinding[DataRow](AsEnumerable(table), "Value", "F")
    if i.Name == "u":
      table = DataTable()
      table.Columns.Add("A", str)
      table.Columns.Add("B", str)
      table.Columns.Add("C", str)
      table.Columns.Add("Value", int)
      for a in F:
        for b in F:
          for c in F:
            table.Rows.Add(a, b, c, 1 if len(a) == 1 and a+b == c else 0)
      i.SetBinding[DataRow](AsEnumerable(table), "Value", "A", "B", "C")
    if i.Name == "v":
      table = DataTable()
      table.Columns.Add("A", str)
      table.Columns.Add("B", str)
      table.Columns.Add("C", str)
      table.Columns.Add("D", str)
      table.Columns.Add("Value", int)
      for a in F:
        for b in F:
          for c in F:
            for d in F:
              table.Rows.Add(a, b, c, d, 1 if len(a) == 1 and a+b+c == d else 0)
      i.SetBinding[DataRow](AsEnumerable(table), "Value", "A", "B", "C", "D")
    if i.Name == "positives":
      table = DataTable()
      table.Columns.Add("F", str)
      table.Columns.Add("Value", int)
      for f in F:
        table.Rows.Add(f, 1 if f in S_plus else 0)
      i.SetBinding[DataRow](AsEnumerable(table), "Value", "F")
    if i.Name == "negatives":
      table = DataTable()
      table.Columns.Add("F", str)
      table.Columns.Add("Value", int)
```

```
    for f in F:
      table.Rows.Add(f, 1 if f in S_minus else 0)
      i.SetBinding[DataRow](AsEnumerable(table), "Value", "F")
  if i.Name == "dummy":
    table = DataTable()
    table.Columns.Add("K", int)
    table.Columns.Add("Value", int)
    for j in xrange(k):
      table.Rows.Add(j, 0)
    i.SetBinding[DataRow](AsEnumerable(table), "Value", "K")
solution = context.Solve()
if solution.Quality == SolverQuality.Optimal \
  or solution.Quality == SolverQuality.LocalOptimal:
  for d in solution.Decisions:
    if d.Name == "x":
      x = d.GetValues()
    if d.Name == "y":
      y = d.GetValues()
    if d.Name == "z":
      z = d.GetValues()
  cfg = varsToCFG(x, y, z)
  context.ClearModel()
  return cfg
else:
  context.ClearModel()
  return None
```

We redefine a parser for obtained grammars. Not only does it answer membership queries but also can check whether there are more than one way for a word to be parsed.

```
class Parser(object):
  """A parser class for QGNF grammars"""

  def __init__(self, productions):
    self.__prods = productions
    self.__cache = dict()

  def __parse(self, w, var):
    """A parsing with checking that it is not ambiguous
    Input: a word, a non-terminal (an integer)
    Output: the number of ways of parsing"""
    if (w, var) in self.__cache:
      return self.__cache[w, var]
    else:
      n = len(w)
      if n == 1: return int((var, w) in self.__prods)
      counter = 0
      for p in self.__prods:
        if p[0] == var and p[1] == w[0]:
          if n > 2 and len(p) == 4:
            for i in xrange(2, n):
              cl = self.__parse(w[1:i], p[2])
              cr = self.__parse(w[i:], p[3])
              counter += cl*cr
          elif len(p) == 3:
            counter += self.__parse(w[1:], p[2])
      self.__cache[w, var] = counter
      return counter
```

```
def accepts(self, word):
  """Membership query
  Input: a string
  Output: true or false"""
  self.__cache.clear()
  return self.__parse(word, 0) > 0

def ambiguousParse(self, word):
  """Check if parsing is unambiguous
  Input: a string that is accepted by this parser
  Output: true or false"""
  self.__cache.clear()
  return self.__parse(word, 0) != 1

def grammar(self):
  return self.__prods
```

Now we are in position to give a procedure for using the `synthesize` function in a way that minimizes its running time (by passing to it as little input data as possible) while maximizing a chance that an obtained grammar is unambiguous (by controlling whether all positive words up to a certain length, `NPOS`, are unambiguously parsed).

```
NPOS = 12   # may be arbitrarily changed

def allWords(alphabet, n):
  """Returns Sigma (<= n)
  Input: the set of chars and an integer
  Output: The set of strings"""
  lang = [set() for i in xrange(n+1)]
  lang[0].add('')
  for i in xrange(1, n+1):
    for w in lang[i-1]:
      for a in alphabet:
        lang[i].add(a+w)
  return reduce(set.__or__, lang)

def findGrammar(alphabet, belong_fun):
  """Main function
  Input: alphabet and a function that tells whether
  a given word belongs to a sought language
  Output: a grammar in QGNF (the set of tuples)"""
  all_words = allWords(alphabet, NPOS)
  all_words.remove('')
  X = [w for w in all_words if belong_fun(w)]
  Y = list(all_words - set(X))
  X.sort(key=len)
  Y.sort(key=len)
  k = 1
  Spos, Sneg = {X[0]}, {Y[0]}
  while True:
    print Spos
    print Sneg
    print k
    G = synthesize(Spos, Sneg, k)
    print G, '\n'
    while G == None:
      k += 1
      print Spos
```

```
      print Sneg
      print k
      G = synthesize(Spos, Sneg, k)
      print G, '\n'
  wp, wn = None, None
  p = Parser(G)
  for w in X:
    if not p.accepts(w):
      wp = w
      break
  for w in Y:
    if p.accepts(w):
      wn = w
      break
  if wp == None and wn == None:
    more_search = False
    for w in X:
      if p.ambiguousParse(w):
        Spos.add(w)
        more_search = True
        break
    if not more_search:
      break
  elif wp == None:
    Sneg.add(wn)
  elif wn == None:
    Spos.add(wp)
  else:
    if len(wp) <= len(wn):
      Spos.add(wp)
    else:
      Sneg.add(wn)
```

The findGrammar function starts from a small sample $S = $ (Spos, Sneg) and $k = 1$. Next, a non-linear program (the synthesize function) is run. Every time it turns out that there exists no solution that satisfies all of the constraints, k is increased by 1. As long as a satisfied grammar is not found, the shortest word in the symmetric difference between the target language and the submitted hypothesis is added to S and a new non-linear program is run again. The whole routine stops after an obtained grammar is consistent with X (all positive words up to a certain length) and Y (all nonempty words over a given alphabet excluding X) and the grammar is unambiguous with respect to X.

The final point we have to do is a bijection between words an routes. In our approach it is convenient to represent it as a function for checking whether a word over {u, r} is associated with a path from $(0, 0)$ point to (n, n) point.

```
def checkWord(w):
  x, y = 0, 0
  for c in w:
    if c == 'r':
      x += 1
    else:
      y += 1
  return x == y
```

The invocation `findGrammar({'r','u'}, checkWord)` begins a main procedure. As a result, the following grammar[2] is obtained:

$$A \rightarrow uB \mid rC \mid uBA \mid rCA$$
$$B \rightarrow r \mid uBB$$
$$C \rightarrow u \mid rCC$$

Employing the methodology specified in Sect. 8.1.2 gives the following set of equations:

$$A(z) = zB(z) + zC(z) + zA(z)B(z) + zA(z)C(z),$$
$$B(z) = z + zB^2(z),$$
$$C(z) = z + zC^2(z),$$

which yields

$$A(z) = \frac{\sqrt{1 - 4z^2} + 4z^2 - 1}{1 - 4z^2}.$$

This is an OGF for the sequence $\langle 0, 0, 2, 0, 6, 0, 20, 0, 70, 0, \ldots \rangle$. However, when $|w| = 2n$ the word w describes a path to (n, n), so we need every second element from the sequence. In addition to this, let us set $a_0 = 1$.

$$\frac{\sqrt{1 - 4z} + 4z - 1}{1 - 4z} + 1 = (1 - 4z)^{-1/2}.$$

From the last expression we can easily extract sought numbers $\langle 1, 2, 6, 20, 70, \ldots \rangle$.

Compositions of a natural number

Problem formulation

Compositions are merely partitions in which the order of summands is considered. For example, there are four compositions of 3: $3, 1+2, 2+1, 1+1+1$. The problem we are dealing with herein is to count the number of compositions of n with no part greater than 3.

Bijection

There is a straightforward one-to-one correspondence between the compositions and a language over the alphabet $\Sigma = \{a, b, c\}$. The sign a is associated with 1, bb with 2, and ccc with 3. So we have, for example, $acccbb$ corresponding with $1 + 3 + 2$, while abc does not belong to the language.

---

[2]It is undecidable whether a CFG is ambiguous in the general case, but in many particular cases, like in the instance above, the unambiguity of a grammar may be proved.

Obtained generating function

This is an OGF for the sequence $\langle 0, 1, 2, 4, 7, 13, 24, 44, 81, 149, \ldots \rangle$:

$$A(z) = \frac{z + z^2 + z^3}{1 - z - z^2 - z^3} \, .$$

Rooted plane trees

Problem formulation

If T is a connected undirected graph without any cycles, then T is called a *tree*. A pair (T, r) consisting of a tree T and a specified vertex r is called a *rooted tree* with *root* r. If, additionally, for each vertex, the children of the vertex are ordered, the result is a *rooted plane tree*. We are to find the number of unlabeled rooted plane trees with n vertexes.

Bijection

Let w be a word formed with the letters x and y. We will say that w is a *Dyck word* if w is either an empty word or a word $xuyv$ where u and v are Dyck words. A word w is a 1-dominated sequence if and only if there exists a prefix of a Dyck word u such that $w = xu$. Let s_n be the number of unlabeled rooted plane trees with n vertexes. To find an OGF for s_n, we can take advantage of a mapping g that maps a 1-dominated word that has n letters x and $n - 1$ letters y to a tree T with n nodes.

Obtained generating function

This is an OGF for the sequence $\langle 0, 1, 1, 2, 5, 14, 42, 132, 429, 1430, \ldots \rangle$:

$$A(z) = \frac{1 - \sqrt{1 - 4z}}{2} \, .$$

Secondary structure of single-stranded tRNAs

Problem formulation

One of the problems considered in chemoinformatics is to count the number of secondary structures of single-stranded tRNAs having a certain size. To this end, the special graph theory of secondary structures has been developed. We call a graph on the set of n labeled points $\{1, 2, \ldots, n\}$ a *secondary structure* if its adjacency matrix $A = (a_{ij})$ has the following three properties:

1. $a_{i,i+1} = 1$ for $1 \leq i \leq n - 1$.
2. For each fixed i, $1 \leq i \leq n$, there is at most one $a_{ij} = 1$ where $j \neq i \pm 1$.
3. $a_{ij} = a_{kl} = 1$, where $i < k < j$, implies $i \leq l \leq j$.

We are to find the number of secondary structures for n points.

Bijection

A secondary structure on n points is represented by a word $w \in \{\text{a}, (,)\}^n$ in which the parentheses are well-balanced, provided that: (1) $a_{ij} = 1$ for $1 \le i < j - 1 < n$ implies $w_i = ($ and $w_j =)$ and (2) $w_k = \text{a}$ for the remaining positions k.

Obtained generating function

This is an OGF for the sequence $\langle 0, 1, 1, 2, 4, 8, 17, 37, 82, 185, 423, \ldots \rangle$:

$$A(z) = \frac{1 - z - z^2 - \left[1 + z(z^3 - 2z^2 - z - 2)\right]^{1/2}}{2z^2}.$$

8.2 Minimizing Boolean Functions

In this section we are going to show how the decomposition-based algorithm for the induction of a CFG can be applied in the designing of digital (electronic) circuits. All the datapath and control structures of a digital device can be represented as boolean functions, which take the general form:

$$F = \Sigma(A, B, \ldots),$$

where $A, B, \ldots$ are boolean variables. Variables may take only the values 0 (zero) and 1 (one). These boolean functions must be converted into logic networks in the most economical way possible. What qualifies as the "most economical way possible" varies, depending on whether the network is built using discrete gates, a programmable logic device with a fixed complement of gates available, or a fully-customized integrated circuit. But in all cases, minimization of boolean functions yields a network with as small a number of gates as possible, and with each gate as simple as possible.

To appreciate the importance of minimization, consider the two networks in Figs. 8.1 and 8.2. Both behave exactly the same way except for the input combination $(0, 0, 0)$. No matter what pattern of ones and zeros we put into A, B, and C in Fig. 8.1, the value it produces at F will be exactly matched if we put the same pattern of values into the corresponding inputs in Fig. 8.2. Yet the network in Fig. 8.2 uses far fewer gates, and the gates it uses are simpler (have smaller fan-ins) than the gates in Fig. 8.1. Clearly, the minimized circuit should be less expensive to build than the unminimized version. Although it is not true for Fig. 8.2, it is often the case that minimized networks will be faster (have fewer propagation delays) than unminimized networks.

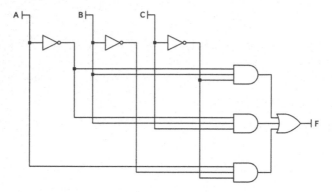

Fig. 8.1 A logic network

Fig. 8.2 A simplified logic
network

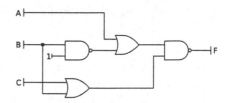

8.2.1   Background and Terminology

The variables in the expression on the right side of a boolean equation are the input
wires to a logic network. The left side of a boolean equation is the output wire of the
network. A *literal* is a variable that is either complemented or not.

Any boolean equation or combinational logic network can be completely and
exactly characterized by a *truth table*. A truth table lists every possible combination
of values of the input variables, and the corresponding output value of the function for
each combination. There are 2^n rows in the truth table for a function or network with
n input variables, so it is not always practical to write out an entire truth table. But
for relatively small values of n, a truth table provides a convenient way to describe
the function or network's behavior exactly.

It is not always necessary to fill in the complete truth table for some real-world
problems. We may have a choice to not fill in the complete table. Both logic networks
depicted in Figs. 8.1 and 8.2 use the function with a truth Table 8.1 in which the two
input combinations: (0, 0, 0) and (1, 1, 1) are *don't cares* (denoted as the x on output),
that is, we do not care what output our logic network produces for these don't cares.

Every row of a truth table with 1 in the output column is called a *minterm*. A convenient way to represent a truth table is a boolean function in *disjunctive normal
form* (DNF). In this form, the function is written as the logical 'or' (indicated by +)
of a set of 'and' terms, one per minterm.[3] For example, the disjunctive normal form

---

[3]In general, an 'and' term of a boolean function in a DNF is allowed to have fewer than n literals.

Table 8.1 An exemplary truth table

A	B	C	F
0	0	0	x
0	0	1	0
0	1	0	1
0	1	1	1
1	0	0	1
1	0	1	0
1	1	0	0
1	1	1	x

for our sample function (treating x as the value 0) would be:

$$F = A'BC' + A'BC + AB'C'.$$

There is also a *conjunctive normal form* (CNF), which represents an expression as a product of sums rather than as a sum of products. However, the material presented further on in this section deals only with disjunctive normal forms. The minterms in our sample function have a total of six literals: $A, A', B, B', C,$ and C'. The network in Fig. 8.2 uses only three literals because no inverter has been used. In the disjunctive normal form of a function, each product term has one literal for each variable.

An *implicant* of a boolean function F is a product term that will generate ones only for combinations of inputs for which F also generates ones. If an implicant P cannot be further reduced—that is, the removal of any literal from P results in a non-implicant—then P is called a *prime implicant* of F. In the example above, $A'B$ is a prime implicant of F.

Figure 8.1 implements our sample function, and demonstrates translating a disjunctive normal form function directly into a logic network. The translation procedure is as follows:

1. Use inverters to generate all possible literals (the six vertical wires on the left in Fig. 8.1).
2. Draw an 'and' gate for each minterm. The fan-in (number of inputs) of each 'and' gate is equal to the number of input variables.
3. Connect the outputs of all the 'and' gates to a single 'or' gate.
4. Connect the inputs of each 'and' gate to a pattern of literals in such a way that it will generate a 1 when the pattern of input values matches the particular minterm assigned to it.

Minimizing boolean functions is very important, because it helps to build simpler electronic circuits. The function

$$F = A'BC' + A'BC + AB'C' + A'B'C'$$

can be formulated as the sum of two prime implicants, $A'B + B'C'$, which can be rewritten as $((A+B')(B+C))'$. The last form of F is directly implemented as a logic network shown in Fig. 8.2. The following theorem reveals that such simplifications are computationally expensive.

Theorem 8.1 *Let $U = \{u_1, u_2, \ldots, u_n\}$ be the set of variables, let $A \subseteq \{0, 1\}^n$ be the set of "truth assignments," and let K be a positive integer. Determining whether there is a disjunctive normal form expression E over U, having no more than K disjuncts, such that E is true for precisely those truth assignments in A, and no others, is NP-complete.*

8.2.2   The Algorithm

We propose here a heuristic method that for a given (not necessarily complete) truth table T, generates a boolean function in DNF with a small number of implicants that cover all minterms in T. Let S_+ be a set of words over $\{0, 1\}$ representing input for ones, while S_- representing input for zeros. The algorithm consists of four steps:

1. Infer a context free grammar $G = \{\{V_0, V_1, \ldots, V_m\}, \{0, 1\}, P, V_0\}$ in CNF from a sample $S = (S_+, S_-)$. Out of three algorithms presented in this book, from Sects. 3.4, 5.3 and 6.2, only the last one works in reasonable time for large input.
2. Divide all non-terminals into four categories: (i) for a non-terminal N we have both rules $N \to 0$ and $N \to 1$ in G, (ii) we have the $N \to 0$ rule and we do not have the $N \to 1$ rule, (iii) we have the $N \to 1$ rule and we do not have the $N \to 0$ rule, and (iv) we do not have either the $N \to 0$ or the $N \to 1$ rule.
3. Generate all sentential forms $\Omega \subseteq \{V_0, V_1, \ldots, V_m\}^n$ that can be derived from V_0. From every $\omega \in \Omega$ free of the fourth category non-terminals make an implicant i in the following way. Let $\omega = ABC\ldots$; for $V = A, B, C, \ldots$, if V is the first category then skip it, if V is the second category then append V' to i, finally, if V is the third category then append V to i. In this way we got a set I of implicants.
4. Using the absorption law, i.e. $x + xy = x$, remove from I all redundant implicants and return the sum of remained implicants.

For the large number of boolean variables the size of counter-examples (S_-) might be so huge that it was impossible to store it in a computer memory. Then, instead of an inference function mentioned above, we may apply in the first step of the algorithm the following routine, which takes only examples (S_+) and outputs a non-circular CFG. (A grammar G is said to be non-circular if the right-hand side of every V_i's rule does not contain a non-terminal V_j, where $j \leq i$.)

```
def synthesize(S_plus):
    V = dict()
    P = set()
    T = {"0", "1"}
    idx = [0]
    initialG(frozenset(S_plus), V, P, T, idx)
    return eliminateUnitProductions(P)
```

The running time of the whole method is mainly dominated by the fourth step, which in turn is hard to analyze. For some obtained grammars it may be polynomially bounded (with respect to n), but for others exponentially.

8.2.3   Our Implementation

The main routine is given at the end of the following listing. On its output we get the string representation of the boolean function in DNF.

```python
from itertools import count, izip
from operator import countOf

def i2str(i):
    """Makes the string representation of an implicant
    Input: An implicant i is represented as a tuple of integers from
    the set {0, 1, 3}.  The three means that there is no appropriate
    variable in a product, for example (1, 3, 0, 1) = A C' D.
    Output: a string (for example: A B' C D')"""
    sym = "ABCDEFGHIJKLMNOPQRSTUVWXYZ"
    lst = []
    for (v, pos) in izip(i, count(0)):
        if v != 3:
            lst.append(sym[pos] + ("'" if v == 0 else ""))
    return " ".join(lst)

def allSentences(grammar, length):
    """Generates all sentential forms of a certain length
    that can be derived in a given grammar
    Input: a grammar (the set of tuples) and an integer
    Output: tuples of integers (each integer is a non-terminal)"""
    Z = {(0,)}
    sentences = set()
    visited = set()
    while Z:
        t = Z.pop()
        n = len(t)
        if n < length:
            for i in xrange(n):
                for p in grammar:
                    if p[0] == t[i] and len(p) == 3:
                        w = t[:i] + p[1:] + t[i+1:]
                        if w not in visited:
                            Z.add(w)
                            visited.add(w)
        else:
            sentences.add(t)
    return sentences

def cfg2bf(g, n):
    """Converts a CFG to a boolean function with n variables
    Input: the set of tuples, an integer
    Output: a boolean function as string like this: A B' + C"""
    d = dict()   # for determining the kind of a non-terminal
    for p in g:
        if p[0] not in d:
```

```
    d[p[0]] = set()
  if len(p) == 2:
    d[p[0]].add(p[1])
implicants = set()
for s in allSentences(g, n):
  if all(v in d and d[v] for v in s):
    i = tuple(3 if d[v] == {'0', '1'} \
      else (1 if '1' in d[v] else 0) for v in s)
    implicants.add(i)
sortkey = lambda i: countOf(i, 3)
t = sorted(implicants, key=sortkey)
ino = len(implicants)
for j in xrange(ino-1):
  for k in xrange(j+1, ino):
    if all(a == 3 or a == b for (a, b) in zip(t[k], t[j])):
      implicants.remove(t[j])
      break
return " + ".join(map(i2str, implicants))
```

8.2.4 Examples

For a small number of variables

Let us minimize the following boolean function:

$$F = AB'C'D'E' + AB'C'D'E + A'B'CD'E' + A'BC'DE' + A'BC'D'E'$$
$$+A'BC'D'E + A'B'C'D'E + A'B'C'D'E' + A'B'C'DE + A'B'CD'E$$
$$+AB'C'DE' + AB'CD'E + AB'CD'E'.$$

We can easily exploit the above presented functions and the algorithm from Sect. 6.2:

```
>>> X1 = set(['10000','10001','00100','01010','01000',
...'01001','00001','00000','00010','00101','10010',
...'10101','10100'])
>>> X0 = set(['01110','01111','11001','11000','01101',
...'01100','11111','11110','11100','11101','11010',
...'11011','00111','00110','01011','00011','10011',
...'10110','10111'])
>>> g = synthesize(X1, X0)
>>> f = cfg2bf(g, 5)
>>> print"F =", f
F = B' C' E' + A' C' D' + B' D' + A' C' E'
```

Fig. 8.3 A simplified logic network built on the basis of the reduced boolean function from the first example

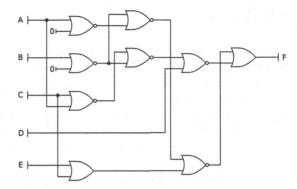

It can be shown that this is the best reduction.[4] A graphical representation of a logic circuit that implements a function F is depicted in Fig. 8.3.

For a large number of variables

Let us consider the set of minterms:

```
>>> X = set([
...'0001010100001010','1011001011001001','0110111101011100',
...'0110000111001001','0001010111001001','0110000101011000',
...'0011000001100000','0110111110111011','0110000101011100',
...'0011000011001001','1011001010111011','0111011001011100',
...'0111011000001010','0011000001011100','1011001001011100',
...'0111011010111011','0011000001011000','0001010101011000',
...'1011001001011000','0111011001100000','1011001001100000',
...'0110000100001010','0011000010111011','0110111111001001',
...'0011000000001010','0110111101011000','0110111101100000',
...'0001010101100000','0111011011001001','0111011001011000',
...'0110000110111011','0001010110111011','1011001000001010',
...'0110000101100000','0001010101011100','0110111100001010'])
```

This time, by means of the above-defined function:

```
>>> g = synthesize(X)
```

and the same function for converting a grammar into a boolean function:

```
>>> f = cfg2bf(g, 16)
```

we obtained 30 implicants (for these 36 words). It is worth to notice that even the famous Espresso, a library for the minimization of two-level covers of boolean functions, does not generate shorter formula in this case.

---

[4]For this purpose (and for building the diagram in Fig. 8.3) we used Logic Friday v. 1.1.4, free software for boolean logic optimization, analysis, and synthesis. Logic Friday is developed at the University of California and it also contains the Espresso logic minimizer, which is the state-of-the-art tool for heuristic logic minimization.

8.3   Use of Induced Star-Free Regular Expressions

The present section is a novel contribution to the field of bioinformatics and combinatorial games by using grammatical inference in the analysis of data. We developed an algorithm for generating star-free regular expressions which turned out to be good recommendation tools, as they are characterized by a relatively high correlation coefficient between the observed and predicted binary classifications. The experiments have been performed for three datasets of amyloidogenic hexapeptides, and for the set of positions in the Toads-and-Frogs game. Our results are compared with those obtained using the current state-of-the-art methods in heuristic automata induction and the support vector machine. The results showed the superior performance of the new GI algorithm on fixed-length samples.

8.3.1   Definitions and an Algorithm

The presented algorithm bases on the concept of decomposition, which was introduced in Chap. 6. However, during the building of the characteristic graph we take advantage of counter-examples as well. As an output, a star-free (i.e., without the Kleene closure operator) regular expression is generated.

Definitions

Definition 8.2 The set of *star-free regular expressions* (SFREs) over Σ will be the set of strings R such that

1. $\emptyset \in R$ which represents the empty set.
2. $\Sigma \subseteq R$, each element a of the alphabet represents language $\{a\}$.
3. If r_A and r_B are SFREs representing languages A and B, respectively, then $(r_A + r_B) \in R$ and $(r_A r_B) \in R$ representing $A \cup B$, AB, respectively, where the symbols $($, $)$, $+$ are not in Σ.

We will freely omit unnecessary parentheses from SFREs assuming that concatenation has higher priority than union. If $r \in R$ represents language A, we will write $L(r) = A$.

Definition 8.3 Let Σ be an alphabet and G be a graph. Suppose that every vertex in G is associated with an ordered pair of nonempty strings over Σ, i.e. $V(G) = \{v_1, v_2, \ldots, v_n\}$ where $v_i = (u_i, w_i) \in \Sigma^+ \times \Sigma^+$ for $1 \leq i \leq n$. Let $C = \{v_{i_1}, v_{i_2}, \ldots, v_{i_m}\}$ be a clique in G. Then

$$r(C) = (u_{i_1} + u_{i_2} + \cdots + u_{i_m})(w_{i_1} + w_{i_2} + \cdots + w_{i_m}) \qquad (8.1)$$

is a star-free regular expression over Σ *induced by* C.

For the sake of simplicity, we also denote the set $L(u_{i_1} + \cdots + u_{i_m}) = \{u_{i_1}, \ldots, u_{i_m}\}$ by U and the set $L(w_{i_1} + \cdots + w_{i_m}) = \{w_{i_1}, \ldots, w_{i_m}\}$ by W in the context of C.

The algorithm

Now, we are going to show how to generate a SFRE compatible with a given sample. These expressions do not have many theoretical properties, but they have marvelous accomplishment in the analysis of some data in terms of classification quality.

Let $S = (S_+, S_-)$ be a sample over Σ in which every string is at least of length 2. Construct the graph G with vertex set

$$V(G) = \bigcup_{s \in S_+} \{(u, w) \mid s = uw \text{ and } u, w \in \Sigma^+\} \qquad (8.2)$$

and with edge set $E(G)$ given by:

$$\{(u, w), (x, y)\} \in E(G) \iff |u| = |x| \text{ and } uy \notin S_- \text{ and } xw \notin S_- . \qquad (8.3)$$

Next, find a set of cliques $\mathscr{C} = \{C_1, C_2, \ldots, C_k\}$ in G such that $S_+ \subseteq \sum_{i=1}^{k} r(C_i)$. For this purpose one can take advantage of an algorithm proposed by Tomita et al. (2006) for generating all maximal cliques. Although it takes $O(n3^{n/3})$-time in the worst case for an n-vertex graph, computational experiments described in Sect. 8.3.2 demonstrate that it runs very fast in practice (a few seconds for thousands of vertexes). Finally, return the union of SRFEs induced by all maximal cliques $\mathscr{C}$, i.e. $e = r(C_1) + r(C_2) + \cdots + r(C_k)$.

In order to reduce the computational complexity of the induction, instead of Tomita's algorithm the ensuing randomized procedure could be applied. Consecutive cliques C_i with their catenations $U_i W_i$ are determined until $S_+ \subseteq \bigcup_{i=1}^{k} U_i W_i$. The catenations emerge in the following manner. In step $i + 1$, a vertex $v_{s_1} = (u, w) \in V(G)$ for which $uw \notin \bigcup_{m=1}^{i} U_m W_m$ is chosen at random. Let $U_{i+1} = \{u\}$ and $W_{i+1} = \{w\}$. Then sets U_{i+1} and W_{i+1} are updated by adding words from the randomly chosen neighbor of v_{s_1}, say v_{s_2}, and subsequently by adding words from the randomly chosen neighbor v_{s_3} of $\{v_{s_1}, v_{s_2}\}$, etc. In the end, a maximal clique C_{i+1} is obtained for which $L(r(C_{i+1})) = U_{i+1} W_{i+1}$. Naturally, $e = r(C_1) + r(C_2) + \cdots + r(C_k)$ fulfills $S_+ \subseteq L(e)$, and the whole procedure runs in polynomial time with respect to the input size.

Here are some elementary properties of a resultant expression e and the complexity of the induction algorithm.

(i) $S_- \cap L(e) = \emptyset$ implies from (8.3).

(ii) If all strings in a sample have equal length, let say ℓ, then all strings from $L(e)$ also are of the same length ℓ.

(iii) Let $n = \sum_{s \in S} |s|$. A graph G, based on (8.2) and (8.3) may be constructed in $O(n^3)$ time. Determining a set of cliques $\mathscr{C}$ and corresponding regular expressions $r(C_1), r(C_2), \ldots, r(C_k)$ also takes no more than $O(n^3)$ time, assuming that the graph is represented by adjacency lists. Thus, the overall computational complexity is $O(n^3)$.

Fig. 8.4 A graph G built
from a sample S, according
to definitions (8.2) and (8.3)

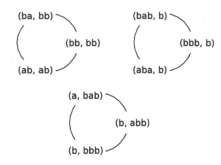

An illustrative run

Suppose $S = (\{bbbb, babb, abab\}, \{bbba, baba, baaa, abaa, aaba, aaab\})$
is a sample (one of possible explanations for the input is: each a follows at least one
b). A constructed graph G is depicted in Fig. 8.4. It has three maximal cliques and
regardless of a method—either Tomita's or randomized algorithm was selected—all
of them would be determined in this case. The final SFRE induced by the cliques is:

$$(ab + ba + bb)(bb + ab) + (aba + bbb + bab)(b) + (b + a)(bab + bbb + abb)$$

Among all words of length four over the alphabet $\{a, b\}$ it does not accept: aaaa,
baaa, abaa, bbaa, aaba, baba, abba, bbba, aaab, but accepts: baab, abab,
bbab, aabb, babb, abbb, bbbb.

An implementation

Please notice that we used here selected useful functions described in Appendix B.
We also took advantage of the `pipetools` package: a library that enables function
composition and partial function application. The main function, `synthesize`,
returns a SFRE which is represented by the list of pairs (tuples).

```python
import networkx as nx
from pipetools import *

def pairFromClique(C):
    """Returns a SFRE induced by C
    Input: the list of nodes (every node is the pair of words)
    Output: the pair of sets of words"""
    U = set()
    W = set()
    for (a, b) in C:
        U.add(a)
        W.add(b)
    return (U, W)

def sfreFromCliques(cliques):
    """Finds a SFRE from cliques
    Input: an iterator over lists (cliques)
    Output: a SFRE as the list of pairs of sets of words"""
    result = []
    for C in cliques:
```

```
    (U, W) = pairFromClique(C)
    result.append((U, W))
  return result

def buildGraph(S):
  """Constructs a graph G
  Input: a sample---the pair of sets
  Output: an undirected networkx graph"""
  Sp, Sn = S[0], S[1]
  V = []
  G = nx.Graph()
  for x in Sp:
    for i in xrange(1, len(x)):
      V.append((x[:i], x[i:]))
      G.add_node((x[:i], x[i:]))
  for i in xrange(len(V) - 1):
    for j in xrange(i+1, len(V)):
      w1 = V[i][0] + V[j][1]
      w2 = V[j][0] + V[i][1]
      if (len(V[i][0]) == len(V[j][0])) \
         and (w1 not in Sn) and (w2 not in Sn):
        G.add_edge(V[i], V[j])
  return G

def accepts(e, x):
  """A membership query
  Input: a SFRE and a word
  Output: true or false"""
  for (p, s) in ((x[:i], x[i:]) for i in xrange(1, len(x))):
    for (L, R) in e:
      if p in L and s in R:
        return True
  return False

synthesize = (pipe
| buildGraph
| nx.find_cliques
| sfreFromCliques)
```

8.3.2   An Application in Classification of Amyloidogenic Hexapeptides

We applied our method to a real bio-informatics task, i.e. classification of amyloidogenic hexapeptides. Amyloids are proteins capable of forming fibrils instead of the functional structure of a protein, and are responsible for a group of diseases called amyloidosis, such as Alzheimers, Huntingtons disease, and type II diabetes. Furthermore, it is believed that short segments of proteins, like hexapeptides consisting of 6-residue fragments, can be responsible for amyloidogenic properties. Since it is not possible to experimentally test all such sequences, several computational tools for predicting amyloid chains have emerged, inter alia, based on physicochemical properties or using machine learning approach.

Datasets

The algorithm for generating star-free regular expressions has been tested over three recently published Hexpepset datasets, i.e. Waltz, WALTZ-DB, and exPafig. The first two databases consist of only experimentally asserted amyloid sequences. Note that the choice of experimentally verified short peptides is very limited since very few data is available. The Waltz dataset was published in 2010 and is composed of 116 hexapeptides known to induce amyloidosis (S_+) and by 161 hexapeptides that do not induce amyloidosis (S_-). The WALTZ-DB has been prepared by the same science team in the Switch Lab from KULeuven, and published in 2015. This dataset expands the Waltz set to total number of hexapeptides of 1089. According to Beerten et al. (2015), additional 720 hexapeptides were derived from 63 different proteins and combined with 89 peptides taken from the literature. In the WALTZ-DB database, 244 hexapeptides are regarded as positive for amyloid formation (S_+), and 845 hexapeptides as negative for amyloid formation (S_-).

SFRE algorithm was also validated and trained on database (denoted by exPafig), which was computationally obtained with Pafig method, and then statistically processed. The exPafig consists of 150 amyloid positive hexapeptides (S_+), and 2259 negative hexapeptides (S_-). As seen, the last database is strongly imbalanced.

Comparative methods

To test the performance of our SFRE approach, the following six additional programs have been used in experiments: the implementation of the Trakhtenbrot-Barzdin state merging algorithm, as described by Lang (1992); the implementation of Rodney Price's Abbadingo winning idea of evidence-driven state merging; a program based on the Rlb state merging algorithm; ADIOS (for Automatic DIstillation Of Structure)—a context-free grammar learning system, which relies on a statistical method for pattern extraction and on structured generalization; our previous approach with directed acyclic word graphs (see Chap. 7), and, as an instance of ML methods, the support vector machine.

Results

Comparative analysis of three measures of merit (sensitivity, specificity, and MCC) is summarized in Table 8.2. These quantities are reported for seven compared predictors and three databases (Waltz, WALTZ-DB, and exPafig). Numerical results of 10×10 cv average-over-fold reported in Table 8.2 show that SFRE has the highest average MCC (0.40) followed by SVM (0.31), ADIOS and Traxbar (0.25), Blue-fringe (0.22), DAWG and Rlb (0.19). Furthermore, SFRE has the highest MCC score compared to the other predictors on each dataset (0.37, 0.38, and 0.44, resp.). Although the results of MCC score seem to be not high (at the level of 0.40), it should be noted that many of the amyloid predictors are reported to have similar or lower values. It is also worth mentioning, that all methods have gained the highest MCC values for the computationally generated exPafig dataset.

SFRE has a higher specificity score than other methods except SVM in case of WALTZ-DB (0.95 to 0.98, resp.) and exPafig databases (both SPC of 1.00). These

Table 8.2 Performance of compared methods on Waltz, WALTZ-DB, and exPafig databases in terms of sensitivity (TPR), specificity (SPC), and Matthews correlation coefficient (MCC)

Method	Waltz			WALTZ-DB			exPafig			
	TPR	SPC	MCC	TPR	SPC	MCC	TPR	SPC	MCC	Avg MCC
SFRE	0.30	0.97	0.37	0.33	0.95	0.38	0.25	1.00	0.44	0.40
SVM	0.35	0.90	0.30	0.15	0.98	0.24	0.22	1.00	0.40	0.31
ADIOS	0.36	0.82	0.22	0.64	0.59	0.20	0.51	0.90	0.34	0.25
Traxbar	0.56	0.61	0.17	0.46	0.76	0.20	0.42	0.96	0.37	0.25
Blue-fringe	0.58	0.53	0.11	0.36	0.85	0.23	0.33	0.96	0.32	0.22
DAWG	0.90	0.13	0.04	0.81	0.47	0.24	0.73	0.80	0.30	0.19
Rlb	0.36	0.70	0.07	0.26	0.90	0.22	0.25	0.97	0.29	0.19

two predictors have a very good capacity at predicting non-amyloid hexapeptides, with SPC higher than 0.90 for each database. The counterpart is their poor sensitivity (true positive rate). Concerning TPR score, DAWG, our proposal from Chap. 7, has the highest value on each database (0.90, 0.81, and 0.73, resp.). SFRE algorithm showed a low sensitivity for each tested dataset (0.30, 0.33, and 0.25, resp.).

The evaluation of SFRE on three amyloidogenic hexapeptide datasets revealed its accuracy to predict non-amyloid segments. We showed that the new grammatical inference algorithm gives the best Matthews correlation coefficient in comparison to six other methods, including support vector machine.

8.3.3 An Application in the Construction of Opening Books

Star-free regular expressions can be applied to build opening books for a selected combinatorial game (Toads-and-Frogs). Many games, including combinatorial ones, have their opening, middle game, and endgame stages. Computer science methods—like MCTS, the minimax algorithm with *alpha-beta* pruning, or the depth-first Proof-Number Search algorithm—work well in the middle and final stages of a game. Usually there are no good methods for controlling the beginning of a game, and that is why opening books, by which we mean stored initial moves, draw much attention from researchers. An opening book, which in essence is a database in one of many diverse forms, can be built by hand with the assistance of human experts or automatically. As regards automatic opening book construction, there are two approaches. In the first approach, games are collected and, after analysis, used to update or extend the opening book. The second approach does not depend on the availability of previously played games. The current opening book is analyzed and extended by means of a heuristic evaluation in an iterative process. In the present section, the first approach to automatic opening book construction has been applied, but with the assumption that the number of available approved games is very limited.

Monte-Carlo Tree Search

Monte-Carlo Tree Search is useful for finding the best move in a game. Possible moves are aggregated in a search tree and a number of random simulations are used to evaluate the long-term potential of each move. The Monte-Carlo Tree Search method has four steps: (a) starting at the root node of the tree, select optimal child nodes until a leaf node is reached; (b) expand the leaf node and choose one of its children; (c) play a simulated game starting with that node; (d) use the results of that simulated game to update the node and its ancestors.

In the experiments we used a program based on an algorithm that differs from the classic Monte-Carlo Tree Search in two fundamental ways. Instead of creating a tree online, steps (a) and (b), a fixed array is used. For updating nodes, step (d), the simple minimax algorithm is applied, but it is executed through a parallel for-loop. Both of the alterations are aimed at dealing with severe time constraints and are not recommended when there is enough time for calculation of the next move.

The Game of Toads-and-Frogs

Toads-and-Frogs is an abstract, two-player, strategy game invented by Richard Kenneth Guy. It is played on a $1 \times n$ strip of squares (a board), initially filled in a certain pattern with toads and frogs pieces with a number of unoccupied places. Toads-and-Frogs players take turns; on his turn, a player may either move one square or jump over an opposing piece onto an empty square. Toads move only eastward, frogs only to the west. Naturally, from the rightmost square, toads cannot move, nor can frogs move from the leftmost square. The first player who is unable to move loses. Although the game may start at any configuration, it is customary to begin with toads occupying consecutive squares on the leftmost end and frogs occupying consecutive squares on the rightmost end of the strip.

Determining the value of an arbitrary Toads-and-Frogs position is NP-hard, which was proved by Jesse Hull in 2000.

A dataset

Games that are taken as a base for opening book construction usually come from human experts' records. Toads-and-Frogs, however, is not so popular and we were unable to find any such record, so we decided to perform a self-play match in order to get 1 000 games. All games started from the position t^{14}ofotofotof^{14}, where t is for toads, f is for frogs, and o is for the empty places. Positions appearing after every move up to the 10th move have been labeled either winning, losing, or neutral. Winning positions are those from which there was always victory. Losing positions are those from which the opponent always defeated us (by 'us' we mean one settled side). Neutral are all the remaining positions. In the construction of a sample for positive words, S_+, all winning positions, while for negative words, S_-, all losing and neutral positions have been chosen. In this way 204 words have been collected,

Table 8.3 The average values of precision and MCC from 5×2 cross-validation tests and their t statistics

The average of a measure			t statistic	
SFRE	ADIOS	Traxbar	ADIOS	Traxbar
P = 0.596	0.274	0.506	$t = 1.504$	0.283
MCC = 0.390	0.000	0.251	$t = 4.795$	0.279

among which 70 were positive. Obtaining regular expression accepts 1 015 words, which constituted the opening book.

Statistical analysis

Let us start with the comparison of the classification performance of three algorithms: ADIOS, Traxbar, and our regular expression based procedure. We used the 5×2 cv t test as described in Sect. 1.2.3. For classification, especially for two-class problems, a variety of measures have been proposed. As our experiments will lie in the combinatorial games context, precision (P) is regarded as primary score, as the goal of classification is to identify more relevant cases than irrelevant ones, next, the Matthews correlation coefficient (MCC), because relying only on precision—without calculating any general measure of the quality of binary classifications—would have been misleading.

The results from SFRE versus ADIOS and SFRE versus Traxbar 5×2 cv[5] tests are gathered in Table 8.3. The averages of both measures are better (larger) for SFRE, although in the case of SFRE versus Traxbar, t statistics are outside the interval $(-t_{\alpha/2, 5}, t_{\alpha/2, 5})$, for relatively high $\alpha/2 = 0.4$ ($t_{0.4, 5} = 0.265$). Thus, the test rejects the hypothesis that the two classification algorithms have the same precision and MCC at significance level $\alpha/2 = 0.4$ for SFRE versus Traxbar and at significance level $\alpha/2 = 0.1$ for SFRE versus ADIOS ($t_{0.1, 5} = 1.475$). This is a strong rationale behind using our algorithm to build an opening book from given data.

In order to check whether using the opening book for the first 10 moves is profitable, the following experiment and binomial test have been done. We played 200 games against an MCTS program without an opening book and counted the number of wins. The side, i.e., playing toads or frogs, was chosen randomly before every game. Using the opening book eliminated the need for the modified program to calculate the next move unless none of the available positions was stored in it. The situation of multiple candidate moves was resolved by random choice. For $i = 1, 2, \ldots, 200$, let z_i be a 0/1 Bernoulli random variable that takes the value 1 when in ith game the modified program wins and 0 when it loses. If there is no improvement by using the opening book, then the probability p of a win is less than or equal to $1/2$. The binomial random variable Z denotes the total number of wins:

---

[5]Because we deal with an imbalanced data set ($|S_+| = 70, |S_-| = 134$), stratified 2-fold was applied. Stratified k-fold is a variation of k-fold that returns stratified folds: Each set contains approximately the same percentage of samples of each target class as the complete set.

$$Z = \sum_{i=1}^{200} z_i \ .$$

We would like to find the critical value e for which $P\{Z \geq e\} < \alpha$. Since

$$P\{Z = j\} = \binom{200}{j} p^j (1 - p)^{200-j} \ ,$$

one can easily determine that $e = 113$ (for $p = 1/2$ and $\alpha = 0.05$). It is reasonable to reject the hypothesis that $p \leq 1/2$ if we see $Z \geq e$. Because the number of wins was exactly 113, we can claim that our program plays significantly better than its basic version, i.e., $p > 1/2$.

The proposed idea is not free from objections. Among the most serious complications are: (a) uncertainty about the possibility of describing good positions in a game by a regular expression, (b) uncertainty about the total covering of the vital initial moves by a regular expression which has been found, and (c) it is unproven whether the method scales well on two-dimensional games.

8.4 Bibliographical Background

For a more detailed presentation and advanced applications of generating functions, the reader is referred to Bender and Williamson (2006) and Graham et al. (1994). The idea of constructing a bijection between a class of combinatorial objects and the words of a language in order to deduce the generating function of the sequence of some parameter p on these objects is known as the Schützenberger methodology and has been developed since the 1960's. As a good bibliographical starting point, see Delest (1994) and Kuich and Salomaa (1985).

For a deeper discussion of normal forms, we refer the reader to Wood (1970). In a book by Hopcroft et al. (2001) one can find that it is undecidable whether a CFG is ambiguous in the general case. The ambiguity of a context-free grammar might be checked by means of the LR(k) test or other methods Basten (2009).

Three last applications from the Sect. 8.1.3 come from Wieczorek and Nowakowski (2016). Particularly, the special graph theory of secondary structures was developed by Waterman (1978).

Theorem 8.1, that tells us that the logic minimization problem is NP-hard for boolean functions expressed in DNF, was proved by Gimpel (1965). Exact and heuristic algorithms for the simplification of boolean functions and the synthesis of logic circuits are the subject of much debate and have a broad literature: see the research of Nelson et al. (1995) and Buchfuhrer and Umans (2011). The Espresso logic minimizer was developed at IBM by Robert Brayton. Rudell and Sangiovanni-Vincentelli (1987) published the variant Espresso-MV, which has inspired many derivatives. To the best of our knowledge, Sect. 8.2 is the first to present an approach to logic minimization that employs a GI method.

The star-free regular expression approach and its analysis with respect to the classification of amyloidogenic hexapeptidesis are taken from the work by Wieczorek and Unold (2016). Experiment design and statistical analysis given there is performed much thoroughly than in this book. The three Hexpepset datasets, i.e. Waltz, WALTZ-DB, and exPafig, are reported by Serrano et al. (2010), Beerten et al. (2015), and Wieczorek and Unold (2014), respectively. Comparative methods are described by: Trakhtenbrot and Barzdin (1973) and Lang (1992) (Traxbar), Lang et al. (1998) (Blue-fringe), Lang (1997) (Rlb), Solan et al. (2005) (ADIOS), Cortes and Vapnik (1995) (SVM).

The application of the SFRE method in a combinatorial game, Toads-and-Frogs, was reported by Wieczorek and Nowakowski (2015). The basic program (without an opening book) was introduced by Wieczorek et al. (2013). The general approach to opening books was discussed by Hyatt (1999) and Lincke (2001). A good bibliographical study on the MCTS topics is made by Browne et al. (2012). As a source of information on the Toads-and-Frogs game, especially in the context of its computational complexity and analysis from the perspective of combinatorial game theory, we encourage the reader to read articles by Erickson (1996) and Thanatipanonda (2008).

References

Basten HJS (2009) The usability of ambiguity detection methods for context-free grammars. Elect Notes Theor Comput Sci 238(5):35–46

Beerten J, Van Durme J, Gallardo R, Capriotti E, Serpell L, Rousseau F, Schymkowitz J (2015) Waltz-db: a benchmark database of amyloidogenic hexapeptides. Bioinformatics p btv027

Bender EA, Williamson SG (2006) Foundations of combinatorics with applications. Dover Books on Mathematics Series, Dover Publications

Browne C, Powley E, Whitehouse D, Lucas S, Cowling PI, Rohlfshagen P, Tavener S, Perez D, Samothrakis S, Colton S (2012) A survey of monte carlo tree search methods. IEEE Trans Comput Intell AI Games 4(1):1–43

Buchfuhrer D, Umans C (2011) The complexity of boolean formula minimization. J Comput Syst Sci 77(1):142–153

Cortes C, Vapnik V (1995) Support-vector networks. Mach Learn 20:273–297

Delest M (1994) Algebraic languages: a bridge between combinatorics and computer science. DIMACS: series in. Discrete Math Theor Comput Sci 24:71–87

Erickson J (1996) New toads and frogs results. In: Nowakowski RJ (ed) Games of no chance. Cambridge University Press, pp 299–310

Gimpel JF (1965) A method of producing a Boolean function having an arbitrarily prescribed prime implicant table. IEEE Trans Elect Comput 14:485–488

Graham R, Knuth D, Patashnik O (1994) Concrete mathematics: a foundation for computer science. Addison-Wesley

Hopcroft JE, Motwani R, Ullman JD (2001) Introduction to automata theory, languages, and computation, 2nd edn. Addison-Wesley

Hyatt RM (1999) Book learning—a methodology to tune an opening book automatically. ICCA J 22(1):3–12

Kuich K, Salomaa A (1985) Semirings. Springer, Languages, Automata

Lang KJ (1992) Random DFA's can be approximately learned from sparse uniform examples. In: Proceedings of the fifth annual workshop on computational learning theory, ACM, pp 45–52

Lang KJ (1997) Merge order count. Technical report, NECI

Lang KJ, Pearlmutter BA, Price RA (1998) Results of the abbadingo one DFA learning competition and a new evidence-driven state merging algorithm. In: Proceedings of the 4th international colloquium on grammatical inference. Springer, pp 1–12

Lincke TR (2001) Strategies for the automatic construction of opening books. In: Marsland TA, Frank I (eds) Computers and games. Lecture notes in computer science, vol 2063. Springer, pp 74–86

Maurer-Stroh S, Debulpaep M, Kuemmerer N, Lopez de la Paz M, Martins IC, Reumers J, Morris KL, Copland A, Serpell L, Serrano L et al (2010) Exploring the sequence determinants of amyloid structure using position-specific scoring matrices. Nat Methods 7(3):237–242

Nelson VP, Nagle HT, Carroll BD, Irwin JD (1995) Digital logic circuit analysis and design. Prentice-Hall

Rudell RL, Sangiovanni-Vincentelli A (1987) Multiple-valued minimization for pla optimization. IEEE Trans Comput Aided Design Integr Circuits Syst 6(5):727–750

Solan Z, Horn D, Ruppin E, Edelman S (2005) Unsupervised learning of natural languages. Proc Nat Acad Sci U S A 102(33):11,629–11,634

Thanatipanonda TA (2008) Further hopping with toads and frogs. http://arxiv.org/pdf/0804.0640v1.pdf

Tomita E, Tanaka A, Takahashi H (2006) The worst-case time complexity for generating all maximal cliques and computational experiments. Theor Comput Sci 363(1):28–42

Trakhtenbrot B, Barzdin Y (1973) Finite automata: behavior and synthesis. North-Holland Publishing Company

Waterman MS (1978) Secondary structure of single-stranded nucleic acids. In: Studies on foundations and combinatorics, advances in mathematics supplementary studies, vol 1. Academic Press, pp 167–212

Wieczorek W, Nowakowski A (2015) Grammatical inference for the construction of opening books. In: Second international conference on computer science, computer engineering, and social media, CSCESM 2015, Lodz, Poland, IEEE, pp 19–22. 21–23 Sept 2015

Wieczorek W, Nowakowski A (2016) Grammatical inference in the discovery of generating functions. In: Gruca A, Brachman A, Kozielski S, Czachórski T (eds) Man–machine interactions 4, advances in intelligent systems and computing, vol 391. Springer International Publishing, pp 627–637

Wieczorek W, Unold O (2014) Induction of directed acyclic word graph in a bioinformatics task. JMLR Workshop Conf Proc 34:207–217

Wieczorek W, Unold O (2016) Use of a novel grammatical inference approach in classification of amyloidogenic hexapeptides. Comput Math Methods Med 2016, article ID 1782732

Wieczorek W, Skinderowicz R, Kozak J, Juszczuk P, Nowakowski A (2013) Selected algorithms from the 2013 Toads-and-Frogs blitz tournament. ICGA J 36(4):222–227

Wood D (1970) A generalised normal form theorem for context-free grammars. Comput J 13(3):272–277

Appendix A
A Quick Introduction to Python

A.1 General Information about the Language

Python is an interpreted programming language which we can use both: interactively, by putting commands and getting immediate results, and in a batch mode, by processing the whole program stored in a text file. An open-source implementation of the Python programming language[1] with its documentation can be downloaded from ironpython.net.

It is common to use py as the file name extension for Python source files. To start an interactive session we type ipy in the command line console (please notice that so as to use some modules one is obligated to invoke IronPython with the -X:Frames parameter), whereas to run a program stored in a file name.py we type ipy name.py.

Using Python, programmers are allowed to follow procedural as well as object-oriented programming language paradigms. The structure of a typical program is almost the same as in most programming languages: loading necessary modules (by means of the instruction **import** or **from**), the definition of procedures, functions, classes, and the main computational part. We will not present this in a formal way, because in Python we have broad discretion over the program framework; for example, instructions can be put between subroutines, and module import—in the body of a function. An interpreter executes the code of a program from the top of a file downward to the last line. It is also worth to emphasize the following features that distinguish Python in the realm of programming languages:

- Dynamic type-checking: the type of a variable can not be explicitly shown. It is done at runtime.
- Instructions are not ended with semicolons: every instruction is put into a separate line.
- Indentations are mandatory: they play the role of 'begin' and 'end' markups.

---

[1] The present description is based on IronPython 2.7.

© Springer International Publishing AG 2017

W. Wieczorek, *Grammatical Inference*, Studies in Computational Intelligence 673,
DOI 10.1007/978-3-319-46801-3

Table A.1 Built-in types and data structures

Name	Description	Example
`object`	The base class of all types and classes	`class Mine(object):`
`bool`	Boolean type	`True != False`
`int`	Integers	`x = 2`
`long`	Long numbers (with unbounded precision)	`x = 1009872116134L`
`float`	Floating-point numbers	`x = 2.123`
`complex`	Imaginary numbers	`z = (1+2j)*(0.3-4.1j)`
`str`	Strings	`x = "abc de"`
`unicode`	Unicode strings	`x = u"text"`
`list`	Lists (elements can be of different type)	`x = [1, 2, 3]`
`tuple`	Tuples, i.e., unchanging sequences of data	`x = (1, "a", 3)`
`xrange`	Iterators	`for i in xrange(9):`
`dict`	Dictionaries implemented as hash tables	`s = {'A':19, 'Bc':21}`
`file`	Files	`fs = open("d.txt","r")`
`set`	Mutable sets	`z = set([1.3, 'a'])`
`frozenset`	Immutable sets	`z = frozenset(['a', 2])`

- In most cases a code is independent from operating system and therefore compatible to 32 bit and 64 bit computers and also suitable for Windows, Linux, and Mac.
- The possibility of treating a string as a code and executing it (the **eval** function and the **exec** statement).
- A class can be derived from a built-in type.
- Functions and objects can serve as iterators (the instruction **yield**, the method `__iter__(self)`).

Built-in Types and Data Structures

Apart from the basic types listed in Table A.1, one should remember about the Decimal type, which is available after loading decimal module. It offers decimal floating-point arithmetic and has a user alterable precision which can be as large as needed for a given problem:

```
>>> from decimal import *
>>> getcontext().prec = 30
>>> Decimal(1) / Decimal(3)
Decimal('0.333333333333333333333333333333')
```

In Python, even the variables of such basic types as numbers (e.g., **int**, **float**) are instances of certain classes. However, we use them conventionally, like in other programming languages. But it is a fact that in Python all calls are calls by reference. For example, when we write:

```
>>> num = 12
>>> txt = "Alice"
```

then num is an alias to an object of the **int** class that stores the value 12, and txt is an alias to an object of the **str** class that stores the string 'Alice'. Thus, the two above instructions are equivalent to the following Java code:

```
Integer num = new Integer(12);
String txt = new String("Alice");
```

or the following C++ code:

```
int x = 12; int &num = x;
const string txt("Alice");
```

It is important to distinguish between changing and unchanging types. Among types from Table A.1 the ones that can be modified are: **list, dict, file**, and **set**. Let us see, in this context, how lists and tuples behave:

```
>>> tab = [3, 4, 5]
>>> tab[0] += 10
>>> tab
[13, 4, 5]
>>> k = (3, 4, 5)
>>> k[0] += 10
Traceback (most recent call last):
  File "<stdin>", line 1, in <module>
TypeError: 'tuple' object does not support item assignment
```

Below are presented two assignment instructions which take the same effect as in other programming languages:

```
>>> x = 2
>>> x +- 1
>>> x
3
```

However, the underlying process is not so obvious. First, x is a reference to an int 2, and in the end it is a reference to another object, 3, which has been obtained by executing x += 1. That is why there are two ways of a comparison: value-based, by means of the operator ==, and identity-based, by means of the keyword **is**. Two variables are identical if they refer to the same object:

```
>>> a = {1}
>>> b = set(range(1, 20))
>>> c = set(range(2, 20))
>>> a == b-c
True
>>> a is b-c
False
```

There is another way to check whether two variables refer to the same object. We can use the **id** function that for a given object x returns an integer (or long integer) which is guaranteed to be unique and constant for this object during its lifetime.

For container types, an assignment instruction x = y does not copy data. In order to make x an object that stores the same data as in y, we have to use the proper method (if y has such a method) or the copy() function (deepcopy() if shallow copy is not sufficient) from the copy module:

```
>>> x = {1, 2, 3}
>>> y = x.copy()
>>> from copy import copy
>>> x = [1, 2, 3]
>>> y = copy(x)
```

Now, let us illustrate in the series of simple examples how to use most of above-mentioned types. Simultaneously, selected, helpful functions will be introduced.

Whether an integer number will be kept as **int** or **long** is up to the suffix L. Also a sufficiently large result determines it:

```
>>> a = 24
>>> type(a)
<type 'int'>
>>> b = 13**a
>>> print b
54280077037437051277159 5361
>>> type(b)
<type 'long'>
>>> a = 24L
>>> type(a)
<type 'long'>
```

In Python almost all arithmetic operators are used via symbols well known from other languages (Java, C++, etc.). In the examples we have seen the operator $**$ which denotes raising to power.

While writing complex numbers it is convenient to use the compact syntax *real_part* + *imaginary_part*j. Such a number is represented by the pair of two floating-point numbers:

```
>>> a = 2.0 + 3.0j
>>> dir(a)
['__abs__', '__add__', ... , 'imag', 'real']
>>> a.imag
3.0
```

Notice that the purpose of the **dir** function is to obtain the list of an object's member functions (methods) and attributes. This function can also be used with the name of an module or a type as a parameter.

Each letter of a string can be indexed via square brackets, just like elements of a list. But, remember that strings are immutable, i.e., we are not allowed to change, insert or remove their letters. The full list of string methods can be easily found in Python's documentation (or try **dir(str)**).

The **map** function is one of the standard routines in the functional style of programming. Its first argument is another function, which is going to be applied to every item of an iterable (the second argument). This operation returns the list of results:

```
>>> sentence = "Alice has a cat"
>>> map(len, sentence.split(""))
[5, 3, 1, 3]
```

Python has also other built-in functions that are applicable to the functional style of programming: **filter** (to construct a list from those elements which fulfill a given predicate) and **reduce** (to do the operation $(\cdots(((a_0 + a_1) + a_2) + a_3) + \cdots + a_n)$ for a sequence a, where the operator plus can be substituted by an arbitrary binary function). The member function `split(separator)` of the **str** class makes the list of words in the string, using separator as the delimiter string.

In a single list, its elements might be of different types. Beside the standard method of indexing, Python makes slicing and negative indexes accessible:

```
>>> tab = range(3, 11)
>>> print tab
[3, 4, 5, 6, 7, 8, 9, 10]
>>> tab[0], tab[-1], tab[1:4], tab[5:], tab[2:-2]
(3, 10, [4, 5, 6], [8, 9, 10], [5, 6, 7, 8])
```

In order to create a list, we could apply a list comprehension instead of the **range** function:

```
>>> [i for i in xrange(400) if unichr(i) in u"aeiouy"]
[97, 101, 105, 111, 117, 121]
```

The conditional part of this construction is optional. For example the following:

```
>>> [x*x for x in xrange(200, 207)]
```

will generate the list of the square roots of $200, 201, \ldots, 206$. The difference between **range** and **xrange** is essential. The former returns a list, while the latter returns an object of the **xrange** class, i.e., an iterator. It is recommended to use the **xrange** function whenever possible, because then the whole sequence is not kept in a computer's memory. There are many ways to modify a list: `seq.append(x)` adds an element x to the end of a list `seq`; insertion is performed via the method `insert`; to delete one or more elements the **del** operator is used; in place, stable sorting is accomplished via the method `sort`.

A dictionary is useful for storing the key-value pairs. Its great merit is that it retrieves the value for a given key very quickly—through hashing—regardless of the dictionary's size. The key has to therefore be an object, ob, for which it is possible to apply a built-in function **hash**(ob). There is a number of such objects: all of built-in immutable types, all of user defined classes in case the **id** function generates a hash value, and objects of all classes for which the member functions __hash__() and __eq__() (or __cmp__()) have been defined. The simplest method for creating a dictionary instance is to enumerate all pairs between signs {and} :

```
>>> d = {('A', 4): 'knight', ('B', 6): 'bishop'}
```

The same dictionary could be obtained by means of the following instructions:

```
>>> d = {}
>>> d[('A', 4)] = 'knight'
>>> d[('B', 6)] = 'bishop'
```

or else

```
>>> d = dict(zip([('A', 4), ('B', 6)], ['knight', 'bishop']))
```

With the keys being strings, a dictionary may be defined in an even simpler way:

```
>>> days = dict(Mon=1, Tue=2, Wed=3, Thu=4, Fri=5)
>>> print days
{'Thu': 4, 'Tue': 2, 'Fri': 5, 'Wed': 3, 'Mon': 1}
```

The most frequently repeated operations on a dictionary are: member queries (`key` **in** d or `key` **not in** d) and returning the item of d with key `key`, `d[key]`, or setting `d[key]` to `value`, `d[key]=value`. To apply an operation for every item, the member function `iteritems` will be useful:

```
>>> for (key, val) in d.iteritems():
...     print val, "on", key
...
bishop on ('B', 6)
knight on ('A', 4)
```

The **del** operator works for dictionaries as does for lists:

```
>>> del days['Mon']
>>> print days
{'Thu': 4, 'Tue': 2, 'Fri': 5, 'Wed': 3}
```

A Python's set is as in mathematics a collection of distinct objects. In the language, however, we have two kinds of sets: mutable, **set**, and immutable, **frozenset**. The elements of a set must be hashable, so the requirements concerning them are similar to those of dictionary's keys. The most frequently repeated operations on a set s and an element x are: membership queries (x **in** s or x **not in** s), adding a new element (`s.add(x)`), and removing (`s.remove(x)` or `s.discard(x)`). For two sets u and v we can obtain: their union (u | v), intersection (u & v), difference (u − v), and symmetric difference (u^v). To test subset and superset properties the following operators are used: <=, >=, <, >.

The built-in functions that can be applied for sequential types include: **len**(s) for determining the size of a sequence s, **all**(s) for checking whether all elements of s do evaluate to `True`, **any**(s) for checking whether at least one element of s does evaluate to `True`. It should be remembered that the below mentioned objects are equivalent to Boolean `False`:

- `None`,
- `False`,
- `0, 0L, 0.0, 0j`,
- `'', (), []`,
- an empty dictionary or an empty set,
- the instances of user defined classes if only the member function `__nonzero__()` or `__len()__` has been specified and, naturally, in case it returns zero or `False`.

All other values are equivalent to Boolean `True`.

Basic Instructions, Functions, and Exceptions

Let us begin with an example which will illustrate that combining functional programming style structures, the syntax of Python, and useful built-in types allow for coding algorithms very shortly. An efficient sorting algorithm, quicksort—assuming a list x has pairwise different, comparable elements—can be written as follows[2]:

```
def qsort(x):
  if len(x) <= 1: return x
  else: return qsort([a for a in x if a < x[0]]) + \
             [x[0]] + \
             qsort([a for a in x if a > x[0]])
```

The **if** statement has the following syntax:

```
if expression1:
  section1
elif expression2:
  section2
else:
  section3
```

As we have learned, in Python there are no opening and closing keywords or braces to indicate the block of instructions, and thus indentations are very important in this context. Notice an obligatory colon after the first and next expressions (there can be as many as needed) and after the **else** keyword. Usually the expressions perform some tests. Python provides a set of relational and logical operators:

- for equations and inequalities: $<, <=, ==, !=, >=, >$,
- to test identity: **is, is not,**
- for membership queries: **in, not in,**
- for logical conditions: **not, and, or.**

It is allowed to combine comparisons in a way that resembles a mathematical notation:

```
if 1 <= x <= 10:
  print "yes"
```

As we have seen, Python's **if-elif-else** statement examines conditions (which might be just expressions) and performs a chosen set of operations. In a similar way, Python provides a conditional expression that examines a condition and, based on whether the condition evaluates to true or false, returns one of two values. The conditional expression's format is as follows:

```
trueResult if condition else falseResult
```

There are two types of loops: **while** and **for**.

```
while expression:
instructions
```

---

[2]The sign \ helps to break code across multiple lines.

or else:

```
while expression:
  instructions1
else:
  instructions2
```

Instructions in the **else** section are executed only if repetitive tasks terminate naturally, i.e., the test condition (expression) is false, not with the **break** instruction. Python, by the way, also provides the **continue** statement, used when we want to skip the rest of the current repetition of the loop and re-evaluate the condition for the next round.

The **for** loop has this two forms as well:

```
for variable in iterator:
  instructions1
else:
  instructions2
```

Basically, the iterator is one of the following objects: string, tuple, list, file, dictionary (via one of the functions: iteritems(), iterkeys(), itervalues()). Instead, the object of an arbitrary class can be used, provided that the method __iter__() or __getitem__() has been implemented. Even a function can be an iterator; an appropriate example will be given later in this subsection. The generator of elements is achieved also when in a list comprehension square brackets will be substituted by parentheses, for instance:

```
>>> map( str, ((i, 3) for i in xrange(2)) )
['(0, 3)', '(1, 3)']
```

Three types of functions exist in Python: ordinary, anonymous, and methods. Now we are going to describe the first two types. The member functions are described in the next section, which is devoted to object-oriented programming. Within our programs, functions have a scope that defines the areas within the program where their names have meaning. There are two types of a scope: global and local. Generally speaking, the global scope means that a function (but also a variable) is available in a file in which it has been defined or in any file that imports this file. We deal with the local scope of a function (but also a variable) when the function has been defined inside another function.

The framework of a function is as follows:

```
def nameOfFunction(parameters):
  bodyOfFunction
```

For example:

```
def greetings():
  print "You are welcome."
```

The name of a function is the alias to the object of a 'function' type and as such can be assigned to a variable:

```
f = greetings
f()              # We will see "You are welcome."
```

All parameters are passed by reference, however, if a parameter is of an unchanging type, it can not be modified. Python lets our programs provide default values for parameters and support a varying number of parameters (using * before a parameter), for example:

```
>>> def maxNumber(*a):
...     return max(i for i in a if isinstance(i, int))
...
>>> maxNumber(3, 'a', 4)
4
>>> maxNumber(1, 'bcd')
1
```

When a formal parameter of the form **name is present, it receives a dictionary containing keyword arguments:

```
>>> def description(firstName, **schedule):
...     print firstName, "works in a room",
...     for (day, room) in schedule.iteritems():
...       print room, "("+day+")",
...
>>> description("Adam", Mon=3, Fri=2)
Adam works in a room 3 (Mon) 2 (Fri)
>>> description("Eve", Tue=2, Thu=7, Fri=3)
Eve works in a room 2 (Tue) 7 (Thu) 3 (Fri)
```

To make a function behave like an iterator is straightforward. It is sufficient to place the **yield** statement instead of **return** statement. A good illustration of this phenomenon is the following definition of the infinite Fibonacci sequence:

```
def fib():
  n0, n1 = 0, 1
  yield 0
  while True:
    yield n1
    n0, n1 = n1, n0 + n1
```

Usually we want to generate a finite number of elements. For example, the power set of any set S can be defined in a recursive fashion:

```
def subsets(S):
  if not S:
    yield S
  else:
    elem = S.pop()
    for u in subsets(S):
      yield u
      yield u | {elem}
```

An anonymous function (also function literal or lambda abstraction) is a function definition that is not bound to an identifier. It is constructed by means of the **lambda** keyword, for example:

```
>>> square = lambda x: x*x
```

Such a function cannot include: **if** statements, loops, and the **return** statement. The result of an anonymous function is the value of an expression given after the colon. They are mostly used in combination with mapping or reducing:

```
>>> tab = ['ab', 'cde', 'fg']
>>> reduce(lambda x, y: x + y, tab)
'abcdefg'
```

Exceptions, like in C++ or Java, are programming structures used to handle errors. Their syntax commonly seen in practice is as follows:

```
try:
  instructions
except kindsOfErrors:
  error_handling
```

There are many ways to catch an error. Selecting the proper one is dependent on what errors we want to catch, specified or any:

```
except IndexError: pass
except ValueError, ob: pass
except (IOError, OSError), ob: pass
except: pass
```

In order to indicate that a particular error has occurred, the **raise** command should be used. Below, we give two examples that will help to familiarize with exceptions in Python.

```
S = set(map(str, xrange(5)))
n = raw_input("Give a number:")
try:
  S.remove(n)
except KeyError:
  print "There is no such a number in a set S!"
```

The next example illustrates how to deal with self-defined errors:

```
import math, random

class NegativeArgument(Exception): pass

tab = [1,3,2,4,5]
random.seed()

def squareRoot(x):
  if x < 0:
    raise NegativeArgument, x
  else:
    return math.sqrt(x)

try:
  print squareRoot(random.randint(-2, 2))
  print tab[random.randint(0, 10)]
except (NegativeArgument, IndexError), ob:
  print ob
```

Decorators

A decorator is just a callable that takes a function as an argument and returns a replacement function. Let us start simply and then work our way up to a useful decorator.

```
>>> def wrap(f):
...     def inner():
...         ret = f()
...         return ret + 1
...     return inner
...
>>> def foo():
...     return 1
...
>>> decorated = wrap(foo)
>>> decorated()
2
```

We defined a function named `wrap` that has a single parameter `f`. Inside `wrap` we defined a nested function named `inner`. The inner function will call a function `f`, catching its return value. The value of `f` might be different each time `wrap` is called, but whatever function it is it will be called. Finally `inner` returns the return value of `f()` plus one, and we can see that when we call our returned function stored in `decorated` we get the results of a return value of 2 instead of the original return value 1 we would expect to get by calling `foo`.

We could say that the variable `decorated` is a decorated version of `foo`: it is `foo` plus something. In fact, if we wrote a useful decorator we might want to replace `foo` with the decorated version altogether so we would always get our 'plus something' version of `foo`. We can do that without learning any new syntax simply by re-assigning the variable that contains our function:

```
>>> foo = wrap(foo)
```

Now any calls to `foo()` will not get the original `foo`, they will get our decorated version. Python has provided support to wrap a function in a decorator by pre-pending the function definition with a decorator name and the @ symbol:

```
>>> @wrap
... def foo():
...     return 1
...
>>> foo()
2
```

Let us write a more useful decorator:

```
def memoize(function):
  cache = {}
  def decorated_function(*args):
    if args in cache:
      return cache[args]
    else:
      val = function(*args)
      cache[args] = val
      return val
  return decorated_function
```

This is an implementation of a memoization—an optimization technique used primarily to speed up routines by storing the results of expensive function calls and returning the cached result when the same inputs occur again. Its advantage is especially seen when combining with recursive functions. Often in those cases the technique reduces the computational complexity of a function from exponential to polynomial time. It is sufficient to mention the Fibonacci sequence:

```
@memoize
def fib(n):
    return n if n <= 1 else fib(n-1) + fib(n-2)
```

Without the decorator `memoize` this function would be highly inefficient. Now, it takes only $O(n)$ operations.

A.2   Object-Oriented Programming

We are starting this section with a brief characteristic of object-orienting programming in Python, often recalling the syntax of such standard OO languages as C++ and Java. In Python, every method may be overwritten in a derived class; in the terms of C++ we would say that every method is virtual. There are also other aspects in which Python is analogous to C++ or Java: overloading operators, multiple inheritance, the existence of a class that is the parent of all classes, and garbage collection.

We define a class using the following syntax:

```
class className(base_class):
    body_of_class
```

In the body of a class, its member functions are defined by means of the **def** keyword. Any attribute or method of the form__name (two leading underscores) is a private member, the remaining ones are public. Besides methods and attributes, a class might posses properties, which we will see later in this section.

Creating Objects

The way in which we create objects is very close to that of Java; the only differences are the consequence of the lack of a 'new' operator and type declaration:

```
ob = myClass(args)
```

The role of a constructor plays the method__init__(), which may be omitted if it is unnecessary. Let us introduce the following example:

```
class Polynomial(object):
    def __init__(self, formula):
        self.__formula = formula
    def getFormula(self):
        return self.__formula
```

The first parameter of every (non-static) method has to be `self`, which in this case, like the **this** keyword in C++ or Java, is a name that refers to the object itself, and it is also the same as what is referenced by a name on the outside of the object, such as `p1` below. Every method that is named with two leading and two trailing underscores is called a special method. They are used for overloading operators and adapting a class to exploit it like built-in types. Members are referred to by a dot:

```
>>> p1 = Polynomial("2*x**4 + 7*x - 2")
>>> p1.getFormula()
'2*x**4 + 7*x - 2'
```

The class `Polynomial` has one attribute. Its name is preceded by two underscores, so the attribute is private and any outside reference to it is prohibited:

```
>>> p1.__formula
Traceback (most recent call last):
  File "<stdin>", line 1, in <module>
AttributeError: 'Polynomial' object has no attribute '__formula'
```

Python follows the philosophy of "we are all adults here" with respect to hiding attributes and methods; i.e. one should trust the other programmers who will use its classes. Using plain (it means public, without two underscores) attributes when possible is recommended. An attribute is 'declared' and initialized in any method, simply by associating to it a value. What is more, in Python, adding a new attribute to an existing object is allowed, for example:

```
>>> p1.var ='x'
```

Properties and Special Methods

Let us consider the class representing a polynomial once again:

```
class Polynomial(object):
  def __init__(self, formula, var):
    self.__formula = formula
    self.__var = var
  def getFormula(self):
    return self.__formula
  def getVariable(self):
    return self.__var
  def newVariable(self, newvar):
    self.__formula = self.__formula.replace(self.__var, newvar)
    self.__var = newvar
```

An instance of this class can be modified in the following way:

```
>>> p1 = Polynomial("-2*x**2 + 3*x + 1", "x")
>>> p1.getFormula()
'-2*x**2 + 3*x + 1'
>>> p1.getVariable()
'x'
>>> p1.newVariable('y')
>>> p1.getFormula()
'-2*y**2 + 3*y + 1'
```

It turns out that the objects of `Polynomial` can be exploited more conveniently by adding a property to the class. The **property** function is responsible for this simplification. Consequently, only by means of one name, getting and setting are performed:

```
class Polynomial(object):
  def __init__(self, formula, var):
    self.__formula = formula
    self.__var = var
  def getFormula(self):
    return self.__formula
  def getVariable(self):
    return self.__var
  def newVariable(self, newvar):
    self.__formula = self.__formula.replace(self.__var, newvar)
    self.__var = newvar
  variable = property(fget=getVariable, fset=newVariable)
```

Now, using the property, the following session with Python's interpreter may occur:

```
>>> p1 = Polynomial("-2*x**2 + 3*x + 1", "x")
>>> p1.getFormula()
'-2*x**2 + 3*x + 1'
>>> p1.variable
'x'
>>> p1.variable = 'y'
>>> p1.getFormula()
'-2*y**2 + 3*y + 1'
```

The great impact of the concept of properties on the clarity of code is not in doubt.

By using special methods (some of them are listed in Table A.2) not only can one overload operators, but also exploit objects as built-in complex structures, e.g. containers. In the latter case the most common defined special methods are presented in Table A.3.

Suppose that to the class `Polynomial` the following special method has been added:

```
def __call__(self, num):
  exec(self.__var + "=" + str(num))
  return eval(self.__formula)
```

Then, thanks to overloading the () function call operator, it is possible to write and evaluate the polynomial at a specific value in a natural way:

```
>>> p = Polynomial("x**2 - 2*x + 1", "x")
>>> p(3)
4
```

Static Methods

Every class is a type, not an object, and every instance of a class has its own copy of class attributes. However, sometimes there are occasions when an implementation is

Table A.2 Basic special methods

Name	Description	Example
__init__(self, arg)	Called after the instance has been created	x = X(a)
__call__(self, arg)	Allowing an object to be invoked as if it were an ordinary function	x(a)
__eq__(self, ob)	Returns True if x and y are equal	x == y
__ne__(self, ob)	Returns True if x and y are not equal	x != y
__le__(self, ob)	Returns True if x is less than or equal to y	x <= y
__lt__(self, ob)	Returns True if x is less than y	x < y
__ge__(self, ob)	Returns True if x is greater than or equal to y	x >= y
__gt__(self, ob)	Returns True if x is greater than y	x > y
__nonzero__(self)	Returns True if x is non-zero value	**if** x:
__str__(self)	Computes the 'informal' string representation of an object	**print** x

Table A.3 Selected special methods used in container operations

Name	Description	Example
__contains__(self, x)	Returns True if x belongs to a sequence y or if x is a key in a map y. The method is also responsible for checking the not in query	x **in** y
__len__(self)	Returns the cardinality of y	**len**(y)
__getitem__(self, k)	Returns the k-th element of y or the value for a key k in a map y	y[k]
__setitem__(self, k, v)	Assigns a value v to the k-th element of a sequence y (or for a map as above)	y[k] = v
__delitem__(self, k)	Deletes the k-th element of y or a key k along with its value from a dictionary y	**del** y[k]
__iter__(self)	Returns an iterator for a container y	**for** x **in** y:

more elegant if all instances share certain data. It would be best if such static (shared) data were 'declared' as the part of a class. This idea can be done in Python like in C++ or Java. A static data member is created in the body of a class but beyond the scope of any **def** instruction. A static method, in turn, is preceded by a standard decorator **staticmethod** and there is a lack of the self parameter:

```python
class Value(object):
  ni = 0
  ns = 0
  def __init__(self, x):
    self.x = x
    if isinstance(x, int):
      Value.ni += 1
    elif isinstance(x, str):
      Value.ns += 1
  @staticmethod
  def obCounter():
    return Value.ni + Value.ns
```

Naturally, in order to refer to static data members or static member functions there is no need to create an instance of a class:

```python
>>> Value.obCounter()
0
>>> o1 = Value(3)
>>> o2 = Value("abc")
>>> o3 = Value("d")
>>> Value.obCounter(), Value.ni, Value.ns
(3, 1, 2)
```

Inheritance and Polymorphism

Inheritance, which lets us derive a new class from an existing or base class, is implemented as shown here:

```python
class Derived(Base):
  class_body
```

The syntax of multiple inheritance is as follows:

```python
class Derived(Base1, Base2, Base3):
  class_body
```

There are two ways for invoking the constructor function of a base class: by means of the **super** built-in function or directly as shown in a class C below:

```python
class A(object):
  def __init__(self, a):
    self.a = a

class B(A):
  def __init__(self, a, b):
    super(B, self).__init__(a)
    self.b = b

class C(A):
  def __init__(self, a, c):
    A.__init__(self, a)
    self.c = c
```

In a derived class we are allowed to overload (re-define) any function that is defined in a base class. Therefore, any method can take different forms in derived classes and the task of selecting a proper one is committed to an interpreter (so-called polymorphism).

Although in Python declaring pure-virtual (abstract) functions (like in C++) or interfaces (like in Java) are not allowed, there is nothing to stop us from achieving the same effect:

```
class Person(object):
  def identifier(self):
    raise NotImplementedError, "Person.identifier()"
  def skills(self):
    raise NotImplementedError, "Person.skills()"
```

A.3   Summary

In this appendix, first, the general information about the Python programming language has been presented. Next, the way in which object-oriented paradigm is put to practice in the language has been described. Using the series of examples, the effortlessness of basic and advanced procedural, functional, and OO techniques has been illustrated. Regardless of certain differences between Python and other 'standard' languages like C++, Java, CSharp, etc., the Python language must be included among general-purpose and powerful programming languages.

It is worth remembering—bedside a straightforward syntax—what lies behind the popularity of Python in the computer science community. So this is because of the great number of ready to use modules from many fields. On the page pypi.python.org one can find packages devoted to areas such as: cryptography, data bases, geographical information systems, image processing, networking, parallel processing, XML, evolutionary algorithms, and even game programming. For the engineering and scientific purposes, the NumPy and SciPy (www.scipy.org) libraries have been prepared. They include numerical recipes for manifold domains such as: statistics, algebra, combinatorial and numerical optimization, and signal processing. These tools provide methods for doing numerical integration and differentiation as well. For those who are model their problems as linear programs, non-linear programs, or in the form of more general constraints, a great help has been brought by integrating IronPython with the Optimization Modeling Language (OML) and Microsoft Solver Foundation (MSF). Algorithms for automata may be written compactly using the FaDO package. As far as automata are concerned, there is also COMPAS (Chmiel and Roman 2010), a computing package, that allows to easily operate on synchronizing automata, verifying new synchronizing algorithms etc. (a DFA can be synchronized if there exist a word which sends any state of the automaton to one and the same state). For graph algorithms and manipulation the NetworkX package has been designed and written. In Appendix B we are going to briefly present a few topics.

Appendix B
Python's Tools for Automata, Networks, Genetic Algorithms, and SAT Solving

B.1 FAdo: Tools for Language Models Manipulation

FAdo system is a set of tools for regular languages manipulation. Regular languages can be represented by regular expressions (regexp) or finite automata. Finite automata may be deterministic (DFA) or non-deterministic (NFA). In FAdo these representations are implemented as Python classes. A full documentation of all classes and methods is provided on the web page http://pythonhosted.org/FAdo/. The newest version of this software is at http://fado.dcc.fc.up.pt/software/. In the consecutive subsection we are going to describe the software from the perspective of grammatical inference demands.

Class DFA
The most convenient way to make the DFA class available is to load the following module:

```
>>> from FAdo.fa import *
```

Assume that we are to create a DFA instance based on Fig. B.1. First, we have to add all states to a newly created automaton:

```
>>> aut = DFA()
>>> aut.addState("q0")
0
>>> aut.addState("q1")
1
>>> aut.addState("q2")
2
>>> aut.addState("q3")
3
```

© Springer International Publishing AG 2017   129
W. Wieczorek, *Grammatical Inference*, Studies in Computational Intelligence 673,
DOI 10.1007/978-3-319-46801-3

Fig. B.1 An exemplary
DFA

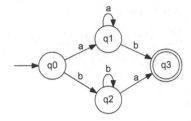

Please notice that the `addState` method returns an index of a just created state. These indexes have to be remembered, because they are needed for inserting transitions and other operations:

```
>>> aut.addTransition(0, 'a', 1)
>>> aut.addTransition(0, 'b', 2)
>>> aut.addTransition(1, 'a', 1)
>>> aut.addTransition(1, 'b', 3)
>>> aut.addTransition(2, 'a', 3)
>>> aut.addTransition(2, 'b', 2)
```

The last thing before the automaton is completed is to define the initial state and final states:

```
>>> aut.setInitial(0)
>>> aut.addFinal(3)
```

Now, we can see its alphabet, states, and the transition function:

```
>>> print aut.Sigma
set(['b', 'a'])
>>> print aut.States, aut.Initial, aut.Final
['q0', 'q1', 'q2', 'q3'] 0 set([3])
>>> print aut.delta
{0: {'b': 2, 'a': 1}, 1: {'b': 3, 'a': 1}, 2: {'b': 2, 'a': 3}}
```

Naturally, any modifications to `aut` should be done via appropriate methods. For example, in order to remove a given state and the transitions related with that state it is sufficient to write:

```
>>> aut.deleteState(1)
```

All indexes and transitions have been adequately recalculated:

```
>>> print aut.States, aut.Initial, aut.Final
['q0', 'q2', 'q3'] 0 set([2])
>>> print aut.delta
{0: {'b': 1}, 1: {'b': 1, 'a': 2}}
```

So as to check whether an automaton accepts a given word we can use the following method:

```
>>> aut.evalWordP("bba")
True
>>> aut.evalWordP("abb")
False
```

However, if we want to query about a word composed of letters that do not belong to the alphabet, first we have to add these letters to the alphabet:

```
>>> aut.addSigma('c')
>>> aut.evalWordP("caa")
False
```

Frequently performed operation on a DFA is transforming it to its minimal form, i.e., reducing the number of states. The method that evaluates the equivalent minimal complete DFA using Hopcroft's algorithm is `minimalHopcroft`.

Given a DFA accepting language L, to enumerate the first m words of L according to the quasi-lexicographic order we use the `EnumDFA` class:

```
>>> g = EnumDFA(aut)
>>> g.enum(5)
>>> g.Words
['ba', 'bba', 'bbba', 'bbbba', 'bbbbba']
```

If there are fewer than or equal to m words, the `enum(m)` method enumerates all words accepted by a given automaton.

Elementary regular languages operations as union (|), intersection (&), concatenation (`concat`), complementation (˜), and reverse (`reversal`) are implemented for DFAs (in fact also for NFAs and regexes). To test if two DFAs are equivalent the operator == can be used.

Class NFA

The NFA class inherits, just as DFA does, all members of an abstract class FA. Thus, NFAs can be built and manipulated in a similar way. There is no distinction between NFAs with and without epsilon-transitions. The most important differences between the NFA and DFA classes are as follows:

- An NFA instance can have multiple initial states; we use the `addInitial` instead of the `setInitial` method.
- To an NFA instance we can add more than one transition for a state and a symbol; see how distinct are deltas in this context:

```
>>> print aut.delta
{0: {'b': 1}, 1: {'b': 1, 'a': 2}}
>>> print aut.toNFA().delta
{0: {'b': set([1])}, 1: {'b': set([1]), 'a': set([2])}}
```

- In order to enumerate words accepted by an NFA instance the `EnumNFA` class is used.

The type of an automaton can be switched from deterministic to non-deterministic or vice versa by means of `toNFA` and `toDFA` methods. The `dup` method duplicates the basic structure into a new NFA (it works also for DFAs and regular expressions).

Class regexp

The most convenient way to make the `regexp` class available is to load the following module:

```
>>> from FAdo.reex import *
```

There are two ways for creating the instance of a regular expression. First, we can simply make it from a string through the `str2regexp` function:

```
>>> e = str2regexp("(a + bc)*")
>>> print e
(a + (b c))*
```

In the second way, one can use specifically prepared classes for the following regular expression preserving operations: concatenation (the `concat` class), union (the `disj` class), and Kleene closure (the `star` class). In addition to this, the `emptyset` and `epsilon` classes represent, respectively, the empty set and the empty word. The above-mentioned expression e can be built alternatively:

```
>>> a = regexp("a")
>>> b = regexp("b")
>>> c = regexp("c")
>>> e2 = star(disj(a, concat(b, c)))
```

Notice that to test equivalence with other regular expressions the `equivalentP` method is used. The operator == is only for checking whether the string representations of two regular expressions are equal:

```
>>> e.equivalentP(disj(e2, epsilon()))
True
>>> e == disj(e2, epsilon())
False
```

Like for DFAs and NFAs, the same function (`evalWordP`) verifies if a word is a member of the language represented by the regular expression.

B.2   NetworkX: Software for Complex Networks

NetworkX is a Python language software package for the creation, manipulation, and study of the structure, dynamics, and functions of complex networks. It has data structures for graphs, digraphs, multigraphs and many standard graph algorithms. The newest version of this software and its documentation is ready to download from https://networkx.github.io/.

Creating a Graph

The most convenient way to make required classes available is to load the following module:

```
>ipy -X:Frames
>>> from networkx import *
```

In order to create an undirected graph we use the `Graph` class and its methods for inserting vertexes and edges:

```
>>> g = Graph()
>>> g.add_node(1)
>>> g.add_nodes_from([2, 3])
>>> g.add_edge(1, 2)
>>> g.add_edges_from([(2, 3), (3, 1)])
```

Naturally, in the collective inserting any iterable container may be used instead of a list. We can examine the characteristics of a graph in the series of queries:

```
>>> g.number_of_nodes()
3
>>> g.number_of_edges()
3
>>> g.nodes()
[1, 2, 3]
>>> g.edges()
[(1, 2), (1, 3), (2, 3)]
>>> g.neighbors(3)
[1, 2]
```

Removing nodes or edges has similar syntax to adding:

```
>>> g.remove_edge(3, 1)
>>> g.edges()
[(1, 2), (2, 3)]
>>> g.remove_nodes_from([1, 2])
>>> g.edges()
[]
>>> g.nodes()
[3]
```

In addition to the methods nodes, edges, and neighbors, iterator versions (e.g. edges_iter) can save us from creating large lists when we are just going to iterate through them anyway.

Please notice that nodes and edges are not specified as NetworkX objects. This leaves us free to use meaningful items as nodes and edges. The most common choices are numbers or strings, but a node can be any hashable object (except None), and an edge can be associated with any object x using the add_edge method with an additional argument: g.add_edge(n1, n2, object=x). We can also safely set the attributes of an edge using subscript notation if the edge already exists:

```
>>> K = complete_graph(4)
>>> K[0][1]['color'] = 'red'
```

Fast direct access to the graph data structure is also possible using the following subscript notations:

```
>>> K[0]
{1: {'color': 'red'}, 2: {}, 3: {}}
>>> K[0][1]
{'color': 'red'}
```

Nodes may have their own attributes as well:

```
>>> K.nodes()
[0, 1, 2, 3]
>>> K.add_node(77, name='Martha', age=30)
>>> K.nodes()
[0, 1, 2, 3, 77]
>>> K.node[1]['txt'] = "Peter"
```

The DiGraph class provides additional methods specific to directed edges, e.g. out_edges, in_degree, predecessors, successors etc. To allow algorithms to work with both classes easily, the directed versions of neighbors and degree are equivalent to successors and the sum of in_degree and out_degree respectively.

There is also available a simpler way for creating graphs:

```
>>> h = Graph([(0, 1), (1, 2), (2, 0)])
>>> d = DiGraph([(0, 1), (1, 2), (2, 0)])
>>> h.edges()
[(0, 1), (0, 2), (1, 2)]
>>> d.edges()
[(0, 1), (1, 2), (2, 0)]
```

but remember the order of two vertexes in every edge matters in directed graphs.

Graph Algorithms

The NetworkX package provides a large number of algorithms, operators, and graph generators. In this book, however, we took advantage of algorithms devoted to cliques. That is why only the following two functions have been presented: (i) the max_clique routine for finding the $O(n/\log^2 n)$ approximation of maximum clique in the worst case, where n is the number of nodes; and (ii) the find_cliques routine for searching for all maximal cliques in a graph.

Assume that g is the instance of the Graph class representing an undirected graph from Fig. B.2. Then, we can use the above-mentioned routines in a straightforward manner:

```
>>> for c in find_cliques(g):
...     print c
...
[0, 2, 4, 1]
[0, 2, 4, 3]
>>> from networkx.algorithms.approximation.clique \
... import max_clique
>>> max_clique(g)
set([0, 3])
```

Fig. B.2 An exemplary graph

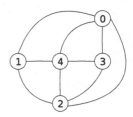

B.3 Pyevolve: A Complete Genetic Algorithm Framework

A genetic algorithm (GA) is a search heuristic that mimics the process of natural selection. This heuristic is routinely used to generate useful solutions to combinatorial optimization problems. Pyevolve was developed to be a complete GA framework written in pure Python. The web page of the package and its documentation is at http://pyevolve.sourceforge.net/.

When a given problem can be solved using a standard form of a chromosome, like 0–1 sequences, one or two dimensional arrays and usually applied operators, then using Pyevolve is extremely easy. Facing a problem for which it is necessary to self-devise the representation of a chromosome and genetic operators, a new class has to be derived from the `GenomeBase` class.

As an illustration let us write a program that randomly generates a weighted undirected graph and then finds a short TSP route in the graph through a simple genetic algorithm:

```python
from networkx import *
from pyevolve import G1DList
from pyevolve import GSimpleGA
from pyevolve import Mutators
from pyevolve import Initializators
from pyevolve import Crossovers
from pyevolve import Consts
import random

random.seed()
G = complete_graph(40)
for (i, j) in G.edges_iter():
  G[i][j]['length'] = random.randint(1, 10000)

def TourInitializator(genome, **args):
  """The initializator for the TSP"""
  global G
  genome.clearList()
  lst = [i for i in xrange(len(G))]
  for i in xrange(len(G)):
    choice = random.choice(lst)
    lst.remove(choice)
    genome.append(choice)

def eval_func(tour):
  """The evaluation function"""
  global G
  total = 0
  num_cities = len(tour)
  for i in xrange(num_cities):
    j = (i+1) % num_cities
    city_i = tour[i]
    city_j = tour[j]
    total += G[city_i][city_j]['length']
  return total

genome = G1DList.G1DList(len(G))
genome.evaluator.set(eval_func)
genome.mutator.set(Mutators.G1DListMutatorSwap)
```

```
genome.crossover.set(Crossovers.G1DListCrossoverOX)
genome.initializator.set(TourInitializator)

ga = GSimpleGA.GSimpleGA(genome)
ga.setGenerations(1000)
ga.setMinimax(Consts.minimaxType["minimize"])
ga.setCrossoverRate(1.0)
ga.setMutationRate(0.05)
ga.setPopulationSize(200)
ga.evolve(freq_stats=50)
best = ga.bestIndividual()
print best
```

B.4   SATisPy: An Interface to SAT Solver Tools

SATisPy is a Python library that aims to be an interface to various SAT (boolean satisfiability) solver applications. Although it supports all solvers as long as they accept the DIMACS CNF SAT format, the easiest way to use SATisPy is via integration with MiniSAT or Lingeling solvers.

A Note about Installation

Now, we describe how to install required software so as to be able to transform some problems into SAT, invoke the MiniSAT solver, and collect the results in Python. Herein is the procedure, assuming that the Windows is an operating system:

1. Go to http://www.cygwin.com/.

   a. Follow the 'install cygwin' link and download setup-x86.exe.
   b. Run the file and follow the instructions to do installation.
   c. The process will create a series of directories rooted at C:\cygwin\.
   d. Add C:\cygwin\bin to the Windows path. Go to the control panel and search for 'path'. Then click on 'edit environment variables for your account'.

2. Go to http://minisat.se/MiniSat.html.

   a. Get the precomplied binary for Cygwin/Windows.
   b. Move this to the C:\cygwin\bin directory and rename the binary file to minisat.exe in this directory.

3. Go to https://github.com/netom/satispy.

    a. Download the current version of SATisPy by clicking the 'Download ZIP' button.
    b. Extract the files to an arbitrary folder.
    c. Go to the subfolder which contains the setup.py file.
    d. Install the package in the Windows command line (in the administrator mode) by `ipy setup.py install`.

4. In the IronPython `\site-packages\satispy` subfolder find the minisat.py file, open it, and change `/dev/null` to `NUL` (which is the Windows counterpart of a virtual place where any data written to it are discarded).

5. Restart the computer.

Now, we can try:

```
>ipy
>>> from satispy import Variable
>>> from satispy.solver import Minisat
>>> v1 = Variable('v1')
>>> v2 = Variable('v2')
>>> v3 = Variable('v3')
>>> exp = v1 & v2 | v3
>>> solver = Minisat()
>>> solution = solver.solve(exp)
>>> if solution.success:
...     print "Found a solution:"
...     print v1, solution[v1]
... print v2, solution[v2]
...     print v3, solution[v3]
... else:
...     print "The expression cannot be satisfied"
...
Found a solution:
v1 False
v2 False
v3 True
```

Expressions do not need to be in conjunctive normal form (CNF). They can be built by creating variables and gluing them together arbitrarily with boolean operators: 'not' -, 'and' &, 'or' |, 'xor'^, and implication >>.

An Illustrative Example

A k-coloring of an undirected graph is labeling of its vertexes with at most k colors such that no two vertexes sharing the same edge have the same color. The problem of generating a k-coloring of a graph $G = (V, E)$ can be reduced to SAT as follows. For every $v \in V$ and every $i \in \{0, 1, \ldots, k - 1\}$, introduce a variable x_{vi}. This variable expresses that a vertex v is assigned the i-th color. Consider the following boolean formulas:

$$\bigvee_{0 \le i < k} x_{vi}, \quad v \in V,$$
$$\neg(x_{vi} \wedge x_{vj}), \quad v \in V, 0 \le i < j < k,$$
$$\neg(x_{vi} \wedge x_{wi}), \quad \{v, w\} \in E, 0 \le i < k.$$

The interpretations satisfying these formulas are in a 1–1 correspondence with k-colorings of (V, E).

An adequate program that uses SATisPy and MiniSAT is presented below:

```python
from satispy import Variable
from satispy.solver import Minisat
from operator import and_, or_
import networkx as nx

N = 20
G = nx.gnm_random_graph(N, 60)
print G.nodes()
print G.edges()

k, x = 6, []
for v in xrange(N):
  x.append([])
  for i in xrange(k):
    x[v].append(Variable(str(v)+','+str(i)))
constraints = []
for v in xrange(N):
  constraints.append(reduce(or_, x[v]))
  for i in xrange(k-1):
    for j in xrange(i+1, k):
      constraints.append(-(x[v][i] & x[v][j]))
  for w in xrange(N):
    if v < w and G.has_edge(v, w):
      for i in xrange(k):
        constraints.append(-(x[v][i] & x[w][i]))
formula = reduce(and_, constraints)

solver = Minisat()
solution = solver.solve(formula)
if solution.success:
  for v in xrange(N):
    for i in xrange(k):
      if solution[x[v][i]]:
        print v, i
else:
  print "A graph G cannot be colored"
```

Appendix C
OML and its Usage in IronPython

C.1 Introduction

In this appendix, the brief description of the Optimization Modeling Language (OML) and the way it is used in IronPython will be presented. We assume that the Microsoft Solver Foundation[3]—which is a set of development tools for mathematical simulation, optimization, and modeling—has been installed in version 3.0 or higher. MSF supports model programming that has linear constraints, unconstrained non-linear programming, and constraint satisfaction programming. Every kind of modeling will be illustrated by an adequate example. In simple terms, given variables $x = (x_1, x_2, \ldots, x_n)$ of possible different domains and constraints in the form of equations and/or inequalities:

$$f_i(x) = 0 \quad i \in I$$
$$f_j(x) \geq 0 \quad j \in J$$

and (optionally) an objective function $c(x)$, we are to choose such an x that satisfies all the constraints and minimizes (or maximizes) the value of $c(x)$. Depending upon the domain of these variables and the form of f and c, we deal with different models. For example, if some or all variables are required to be integers, constraints and an objective function are linear, then a model is known as mixed integer linear programs (MILP).

---

[3] The documentation of MSF tools (and OML as well) is available on the Microsoft web page https://msdn.microsoft.com.

© Springer International Publishing AG 2017
W. Wieczorek, *Grammatical Inference*, Studies in Computational Intelligence 673,
DOI 10.1007/978-3-319-46801-3

C.2   Mathematical Modeling with OML

A model is defined predominantly using linear and non-linear equations or inequalities and, if necessary, using more advanced mathematical functions such as: the sine, arcsine, hyperbolic sine, and the natural exponential function. It is also allowed to use implication in logical expressions. A typical model looks like the following:

```
Model[
   Parameters[...],
   Decisions[...],
   Constraints[...]
   Goals[{Minimize|Maximize}[...]],
]
```

In the decisions section, the variables are declared, then the constraints are written through them. Objectives such as `Maximize` or `Minimize` are specified inside goals. If there are multiple objectives, the model is solved sequentially where the optimal value for the first object is found before considering subsequent objectives. Order does matter for goals, but other sections can be arbitrarily ordered.

Let us consider an example. In the following calculation: go+no=on every letter represents a single digit, but distinct letters represent distinct digits and a letter that is placed in many positions must represent the same digit. The task is to discover an underlying matching. This is the model:

```
Model[
   Decisions[Integers[1, 9], g, n, o],
   Constraints[
      10*g + o + 10*n + o == 10*o + n,
      Unequal[g, n, o]
      ]
]
```

Due to the lack of parameters, the above can be solved by a simple command (`MSFCli.exe`) in an operating system.

C.3   Examples

A Diet Problem—Linear Programming

Consider the problem of diet optimization. There are six different foods: bread, milk, cheese, potato, fish, and yogurt. The cost and nutrition values per unit are displayed in Table C.1. The objective is to find a minimum-cost diet that contains not more than 10 grams of protein, not less than 10 g of carbohydrates, and not less than 8 grams of fat. In addition, the diet should contain at least 0.5 unit of fish and no more than 1 unit of milk.

Because all variables that we are going to define, i.e., the amount of food, are reals, while constraints are linear, we are faced with an LP problem. Below is given a complete IronPython program that contains a suitable OML model.

Table C.1 Cost and nutrition values

	Bread	Milk	Cheese	Potato	Fish	Yogurt
Cost	2.0	3.5	8.0	1.5	11.0	1.0
Protein (g)	4.0	8.0	7.0	1.3	8.0	9.2
Fat (g)	1.0	5.0	9.0	0.1	7.0	1.0
Carbohydrates (g)	15.0	11.7	0.4	22.6	0.0	17.0

```python
import clr
clr.AddReference('Microsoft.Solver.Foundation')
clr.AddReference('System.Data')
clr.AddReference('System.Data.DataSetExtensions')
from Microsoft.SolverFoundation.Services import *
from System.Data import DataTable, DataRow
from System.Data.DataTableExtensions import *
from System.IO import *

# OML Model string
strModel = """Model[
  Parameters[Sets, Food],
  Parameters[Reals, Cost[Food], Protein[Food], Fat[Food],
    Carbohydrates[Food], AtLeast[Food], NoMore[Food]
  ],
  Decisions[Reals[0, Infinity], amount[Food]],
  Constraints[
    Sum[{i, Food}, Protein[i]*amount[i]] <= 10,
    Sum[{i, Food}, Carbohydrates[i]*amount[i]] >= 10,
    Sum[{i, Food}, Fat[i]*amount[i]] >= 8,
    Foreach[{i, Food}, amount[i] >= AtLeast[i]],
    Foreach[{i, Food}, amount[i] <= NoMore[i]]
  ],
  Goals[Minimize[Sum[{i, Food}, Cost[i]*amount[i]]]]
]"""

data = {
  "Cost": [('Bread', 2.0), ('Milk', 3.5), ('Cheese', 8.0),
    ('Potato', 1.5), ('Fish', 11.0), ('Yogurt', 1.0)],
  "Protein": [('Bread', 4.0), ('Milk', 8.0), ('Cheese', 7.0),
    ('Potato', 1.3), ('Fish', 8.0), ('Yogurt', 9.2)],
  "Fat": [('Bread', 1.0), ('Milk', 5.0), ('Cheese', 9.0),
    ('Potato', 0.1), ('Fish', 7.0), ('Yogurt', 1.0)],
  "Carbohydrates": [('Bread', 15.0), ('Milk', 11.7), ('Cheese', 0.4),
    ('Potato', 22.6), ('Fish', 0.0), ('Yogurt', 17.0)],
  "AtLeast": [('Bread', 0.0), ('Milk', 0.0), ('Cheese', 0.0),
    ('Potato', 0.0), ('Fish', 0.5), ('Yogurt', 0.0)],
  "NoMore": [('Bread', 9.9), ('Milk', 1.0), ('Cheese', 9.9),
    ('Potato', 9.9), ('Fish', 9.9), ('Yogurt', 9.9)]
}

context = SolverContext.GetContext();
context.LoadModel(FileFormat.OML, StringReader(strModel));
parameters = context.CurrentModel.Parameters
for i in parameters:
  table = DataTable()
  table.Columns.Add("Food", str)
```

```
table.Columns.Add("Value", float)
for (f, v) in data[i.Name]:
  table.Rows.Add(f, v)
i.SetBinding[DataRow](AsEnumerable(table), "Value", "Food")

solution = context.Solve()
print solution.GetReport()
```

Five parts can be pointed out in this program. The first part contains loading necessary modules. The second part includes the model of a problem written in the OML format. The third part consists of the data representing the particular instance of a problem. They are put directly to a map, but there is nothing to prevent the source of data being in other places (files, data bases, user's answers etc.) The forth part consists of needed objects from the MSF library and binding of values to parameters. In the last part the `Solve` method is invoked and results are reported:

```
===Solver Foundation Service Report===
Date: 31-10-2015 13:14:27
Version: Microsoft Solver Foundation 3.0.1.10599 Enterprise Edition
Model Name: DefaultModel
Capabilities Applied: LP
Solve Time (ms): 199
Total Time (ms): 346
Solve Completion Status: Optimal
Solver Selected: Microsoft.SolverFoundation.Solvers.SimplexSolver
Directives:
Microsoft.SolverFoundation.Services.Directive
Algorithm: Primal
Arithmetic: Hybrid Variables: 6 -> 6 + 4
Rows: 16 -> 4
Nonzeros: 35
Eliminated Slack Variables: 0
Pricing (exact): SteepestEdge
Pricing (double): SteepestEdge
Basis: Slack
Pivot Count: 4
Phase 1 Pivots: 2 + 0
Phase 2 Pivots: 2 + 0
Factorings: 5 + 1
Degenerate Pivots: 0 (0.00 %)
Branches: 0
===Solution Details===
Goals:
goalcef809df_64a6_4b6f_a38c_2fbbc9529f33: 9.17441319845586

Decisions:
amount(Bread): 0
amount(Milk): 0.56440023669306
amount(Cheese): 0.184810786440869
amount(Potato): 0.147017385668799
amount(Fish): 0.5
amount(Yogurt): 0
```

Knapsacking—Integer Programming

The knapsack problem is formulated as follows: given a set of items, each with a weight and a gain, determine which items should be included in a collection so that the total weight is less than or equal to a given limit and the total gain is as large as possible. Here is the solution:

```python
import clr
clr.AddReference('Microsoft.Solver.Foundation')
clr.AddReference('System.Data')
clr.AddReference('System.Data.DataSetExtensions')
from Microsoft.SolverFoundation.Services import *
from System.Data import DataTable, DataRow
from System.Data.DataTableExtensions import *
from System.IO import *
from random import seed, randint

# OML Model string
strModel = """Model[
  Parameters[Sets, Items, dummy],
  Parameters[Reals, Weight[Items], Gain[Items], Limit[dummy]],
  Decisions[Integers[0, 1], x[Items]],
  Constraints[Sum[{i, Items}, Weight[i]*x[i]] <= Limit[0]],
  Goals[Maximize[Profit->Sum[{i, Items}, Gain[i]*x[i]]]]
]"""

seed()
data = {
  "Weight": [(i, float(randint(1, 10000))) for i in xrange(40)],
  "Gain": [(i, float(randint(1, 10000))) for i in xrange(40)],
  "Limit": [(0, 50000.0)]
}
print data

context = SolverContext.GetContext();
context.LoadModel(FileFormat.OML, StringReader(strModel));
parameters = context.CurrentModel.Parameters
for i in parameters:
  if i.Name != "Limit":
    table = DataTable()
    table.Columns.Add("Items", int)
    table.Columns.Add("Value", float)
    for (it, v) in data[i.Name]:
      table.Rows.Add(it, v)
    i.SetBinding[DataRow](AsEnumerable(table), "Value", "Items")
  else:
    table = DataTable()
    table.Columns.Add("dummy", int)
    table.Columns.Add("Value", float)
    for (d, v) in data[i.Name]:
      table.Rows.Add(d, v)
    i.SetBinding[DataRow](AsEnumerable(table), "Value", "dummy")

solution = context.Solve()
for g in solution.Goals:
  print g.Name, "=", g.ToDouble()
for d in solution.Decisions:
  print d.Name
  for v in d.GetValues():
    print int(v[1]), ":", int(v[0])
```

Function minimization—non-linear programming

A nonlinear programming problem (NLP) is an optimization problem where the objective function or some of the constraints are nonlinear. Consider the following problem: minimize $z = 100(y - x^2)^2 + (1 - x)^2$, where $x, y \geq 0$. This is an example of an NLP. It is an NLP because of the multiplication of variables in the objective function. An IronPython solution using OML is as follows:

```
import clr
clr.AddReference('Microsoft.Solver.Foundation')
from Microsoft.SolverFoundation.Services import *
from System.IO import *

strModel = """Model[
  Decisions[Reals[0, Infinity], x, y],
  Goals[Minimize[100*(y - x*x) 2 + (1 - x) 2]]
]"""

context = SolverContext.GetContext();
context.LoadModel(FileFormat.OML, StringReader(strModel));

solution = context.Solve()
print solution.GetReport()
```

A Hamiltonian Cycle—Constraint System

The Hamiltonian cycle problem is a problem of determining whether a cycle in an undirected (or directed) graph, that visits each vertex exactly once, exists. In a program that is given below, the E parameters define the connections between vertexes in an undirected graph, while the variables $\pi(i)$ represent the permutations of $V = \{0, 1, \ldots, N - 1\}$. Thus, a determined cycle goes as follows: $0 \rightarrow \pi(0) \rightarrow \pi(\pi(0)) \rightarrow \cdots \rightarrow 0$.

```
import clr
clr.AddReference('Microsoft.Solver.Foundation')
clr.AddReference('System.Data')
clr.AddReference('System.Data.DataSetExtensions')
from Microsoft.SolverFoundation.Services import *
from System.Data import DataTable, DataRow
from System.Data.DataTableExtensions import *
from System.IO import *
import networkx as nx

strModel = """Model[
  Parameters[Sets, V],
  Parameters[Integers, E[V, V], I[V]],

  Decisions[Integers[0, Infinity], pi[V]],

  Constraints[
    Foreach[{i, V}, pi[i] <= Max[Foreach[{j, V}, I[j]]]],
    Unequal[Foreach[{i, V}, pi[i]]],
    Foreach[{i, V}, {j, V}, (pi[i] == I[j]) -: (E[i, j] == 1)]
  ]
]"""

N = 20
```

```
G = nx.gnm_random_graph(N, N*N/4, directed=True)
print G.nodes()
print G.edges()

context = SolverContext.GetContext();
context.LoadModel(FileFormat.OML, StringReader(strModel));
parameters = context.CurrentModel.Parameters
for i in parameters:
  if i.Name == "E":
    table = DataTable()
    table.Columns.Add("P", int)
    table.Columns.Add("Q", int)
    table.Columns.Add("Value", int)
    for p in xrange(N):
      for q in xrange(N):
        if G.has_edge(p, q):
          table.Rows.Add(p, q, 1)
        else:
          table.Rows.Add(p, q, 0)
    i.SetBinding[DataRow](AsEnumerable(table), "Value", "P", "Q")
  elif i.Name == "I":
    table = DataTable()
    table.Columns.Add("V", int)
    table.Columns.Add("Value", int)
    for v in xrange(N):
      table.Rows.Add(v, v)
    i.SetBinding[DataRow](AsEnumerable(table), "Value", "V")

solution = context.Solve()
if solution.Quality == SolverQuality.Feasible:
  for d in solution.Decisions:
    print d.Name
    for v in d.GetValues():
      print int(v[1]), ":", int(v[0])
else:
  print "There is no Hamiltonian cycle."
```

The last constraint in the OML section is in need of explanation. The $-:$ symbol stands for logical implication, so we have explicitly written there that if in a solution the j-th vertex goes after the i-th vertex then a graph must have the edge (i, j).

Reference

1. Chmiel K, Roman A (2010) COMPAS—a computing package for synchronization. In: Implementation and application of automata, Lecture notes in computer science, vol 6482. Springer, pp 79–86

Printed in the United States
By Bookmasters